Universitext

Advisors

F.W. Gehring
P.R. Halmos
C.C. Moore

C. Godbillon

Dynamical Systems on Surfaces

Translation from the French by
H. G. Helfenstein

With 70 Figures

Springer-Verlag
Berlin Heidelberg New York 1983

Claude Godbillon

Département de Mathématiques, Université Louis Pasteur
7, rue René Descartes, F - 67084 Strasbourg

Original edition "Systèmes dynamiques sur les surfaces"

AMS Subject Classification (1980): 34 C, 58 F

ISBN 3-540-11645-1 Springer-Verlag Berlin Heidelberg New York
ISBN 0-387-11645-1 Springer-Verlag New York Heidelberg Berlin

Library of Congress Cataloging in Publication Data. Godbillon, Claude, 1937-
Dynamical systems on surfaces. (Universitext). Translation of: Systèmes dynamiques sur les surfaces. Bibliography: p. 1. Differentiable dynamical systems. 2. Foliations (Mathematics) I. Title. QA614.8.G6313. 1982. 516'.36. 83-19176
ISBN 0-387-11645-1 (U.S.)

Printed in Germany

Printing and bookbinding: Beltz, Offsetdruck, Hemsbach
2141/3140-543210

Preface

These notes are an elaboration of the first part of a course on foliations which I have given at Strasbourg in 1976 and at Tunis in 1977.

They are concerned mostly with dynamical systems in dimensions one and two, in particular with a view to their applications to foliated manifolds. An important chapter, however, is missing, which would have been dealing with structural stability.

The publication of the French edition was realized by the efforts of the secretariat and the printing office of the Department of Mathematics of Strasbourg. I am deeply grateful to all those who contributed, in particular to Mme. Lambert for typing the manuscript, and to Messrs. Bodo and Christ for its reproduction.

Strasbourg, January 1979.

Table of Contents

I. VECTOR FIELDS ON MANIFOLDS 1

1. Integration of vector fields. 1
2. General theory of orbits. 13
3. Invariant and minimal sets. 18
4. Limit sets. 21
5. Direction fields. 27
A. Vector fields and isotopies. 34

II. THE LOCAL BEHAVIOUR OF VECTOR FIELDS 39

1. Stability and conjugation. 39
2. Linear differential equations. 44
3. Linear differential equations with constant coefficients. 47
4. Linear differential equations with periodic coefficients. 50
5. Variation field of a vector field. 52
6. Behaviour near a singular point. 57
7. Behaviour near a periodic orbit. 59
A. Conjugation of contractions in R. 67

III. PLANAR VECTOR FIELDS 75

1. Limit sets in the plane. 75
2. Periodic orbits. 82
3. Singular points. 90
4. The Poincaré index. 105
5. Planar direction fields. 116
6. Direction fields on cylinders and Moebius strips. 123
A. Singular generic foliations of a disc. 127

IV. DIRECTION FIELDS ON THE TORUS AND HOMEOMORPHISMS OF THE CIRCLE 130

1. Direction fields on the torus. 130
2. Direction fields on a Klein bottle. 137
3. Homeomorphisms of the circle without periodic point. 144
4. Rotation number of Poincaré. 151
5. Conjugation of circle homeomorphisms to rotations. 159
A. Homeomorphism groups of an interval. 166
B. Homeomorphism groups of the circle. 170

V. VECTOR FIELDS ON SURFACES 178

1. Classification of compact surfaces. 178
2. Vector fields on surfaces. 181
3. The index theorem. 188
A. Elements of differential geometry of surfaces. 191

BIBLIOGRAPHY 200

Chapter I. Vector Fields on Manifolds

1. INTEGRATION OF VECTOR FIELDS

Let M be a differentiable manifold without boundary of dimension m and of class C^s, $2 \leqslant s \leqslant +\infty$ (respectively analytic), and let X be a vector field on M of class C^r, $1 \leqslant r \leqslant s-1$ (respectively analytic).

1.1. DEFINITION. An <u>integral curve</u> of X is a map c of class C^1 of an interval J of $\mathbb{R}$ into M satisfying $c'(t) = X(c(t))$ for all $t \in J$.

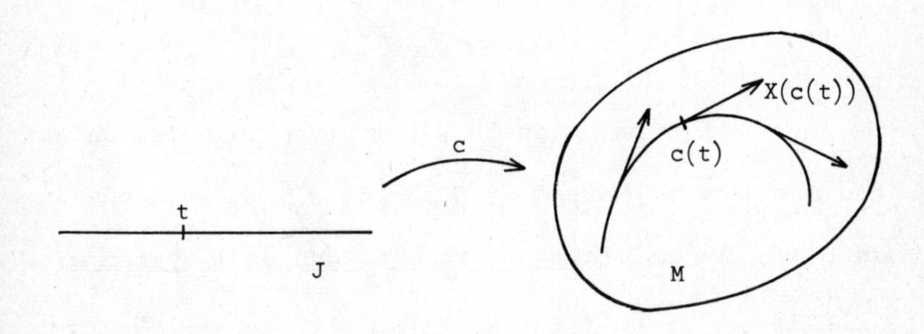

1.2. <u>Examples</u>.

i) If y is a zero of X then the constant map of $\mathbb{R}$ onto y is an integral curve of X.

In this case y is called a <u>singular point</u> of X. A point of M where X does not vanish is a <u>regular point</u>.

ii) If c is an integral curve of X, so is the map $t \mapsto c(t+\tau)$, for all $\tau \in \mathbb{R}$.

iii) If c is an integral curve of X the map $t \mapsto c(-t)$ is an integral curve of the field $-X$.

iv) Let $q: \tilde{M} \to M$ be a covering map. Then the tangent bundle $T(\tilde{M})$ is isomorphic to the inverse image $q^*T(M)$ of the tangent bundle $T(M)$ by the projection q. Hence there is a uniquely defined vector field $\tilde{X}$ on $\tilde{M}$ (in the same differentiability class as X) such that $q^T \circ \tilde{X} = X \circ q$.

In this case the projection $c = q \circ \tilde{c}$ of an integral curve $\tilde{c}$ of $\tilde{X}$ is an integral curve of X.

Conversely every integral curve of X is the projection of an integral curve of $\tilde{X}$.

v) Let X and Y be vector fields on manifolds M and N respectively and $h: M \to N$ a differentiable map satisfying $h^T \circ X = Y \circ h$. Then the image under h of any integral curve of X is an integral curve of Y.

1.3. Remarks.

i) Let $(y_1, \ldots, y_m)$ be a local system of coordinates on an open set U of M, and let X be expressed on U as $\sum_i a_i \frac{\partial}{\partial y_i}$. Then the integral curves of X in U are the solutions of the (autonomous) system of differential equations $y_i' = a_i(y_1, \ldots, y_m)$, $i = 1, \ldots, m$.

ii) If $N = M \times \mathbb{R}$, a vector field Y on N of the form $Z(y,s) + \frac{\partial}{\partial s}$, with $Z(y,s) \in T_yM$, corresponds locally to a non-autonomous system of differential equations $y_i' = b_i(y_1, \ldots, y_m, t)$, $i = 1, \ldots, m$.

The above remark (i) allows a reformulation of the local existence and uniqueness theorems for solutions of differential equations, as well as of the statements about their differentiable dependence on initial conditions as follows (cf. [10], [12]):

1.4. THEOREM. For every point y of M and for every real number τ there exist an open neighbourhood U of y in M,

a number $\varepsilon > 0$,

a map Φ of class C^r (respectively analytic) of $(\tau-\varepsilon, \tau+\varepsilon)\times U$ into M,

satisfying for every point x of U the following properties:

a) $t \to \Phi(t,x)$ is an integral curve of X;

b) $\Phi(\tau, x) = x$;

c) if c is an integral curve of X defined on an interval containing τ and such that $c(\tau) = x$, then $c(t) = \Phi(t,x)$ in a neighbourhood of τ.

Consequences:

i) Two integral curves of X intersecting in a point coincide in a neighbourhood of this point;

ii) let U_i, ε_i, Φ_i, $i = 1,2$, be two sets of data as in theorem 1.4 with the properties a), b), and c). If $\varepsilon = \inf(\varepsilon_1, \varepsilon_2)$ and $V = U_1 \cap U_2$, then $\Phi_1 = \Phi_2$ on $(\tau-\varepsilon, \tau+\varepsilon) \times V$.

1.5. <u>COROLLARY</u>. There exist an open neighbourhood W of $\{0\} \times M$ in $\mathbb{R} \times M$ and a map Φ of class C^r (respectively analytic) of W into M with the following properties satisfied at every point y of M:

a) $\mathbb{R} \times \{y\} \cap W$ is connected;

b) $t \to \Phi(t,y)$ is an integral curve of X;

c) $\Phi(0,y) = y$;

d) if (t',y), $(t+t',y)$ and $(t, \Phi(t',y))$ are in W, then $\Phi(t+t',y) = \Phi(t, \Phi(t',y))$.

Furthermore, if W_i, Φ_i, $i=1,2$, are two such data satisfying a), b), and c), then they also satisfy d), and $\Phi_1 = \Phi_2$ on $W_1 \cap W_2$.

Continuing with these notations we let V be an open set of M such that $\{t\} \times V$ and $\{-t\} \times \Phi(\{t\} \times V)$ are contained in W. Then $\Phi(\{t\} \times V)$ is open in M, and the map $\varphi_t : x \mapsto \Phi(t,x)$ is a diffeomorphism of V onto this open set having the inverse $\varphi_{-t} : z \mapsto \Phi(-t,z)$.

In addition, for $\{t'\} \times V$, $\{t+t'\} \times V$, and $\{t\} \times \Phi(\{t'\} \times V)$ being contained in W as well, we have $\varphi_{t+t'} = \varphi_t \circ \varphi_{t'}$ on V. These remarks and the considerations in §1.7 justify the following definition:

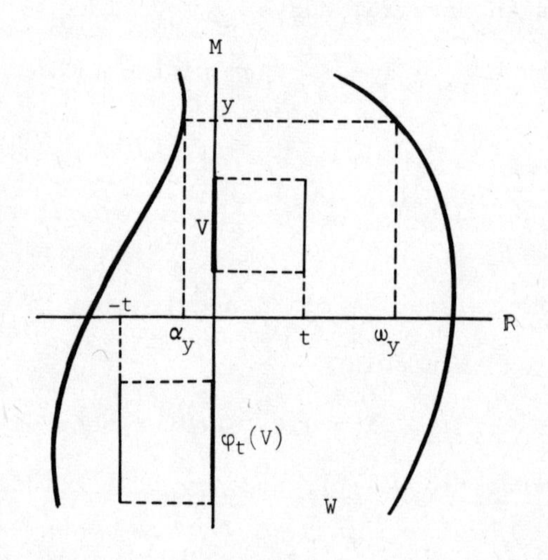

1.6. DEFINITION. A local one-parameter group of diffeomorphisms (or a flow) of class C^r (respectively analytic) of M is an ordered pair (W, Φ), where W is an open neighbourhood of $\{0\} \times M$ in $\mathbb{R} \times M$ and Φ a map of class C^r (analytic) of W into M, having at every point y of M the following properties:

a) $\mathbb{R} \times \{y\} \cap W$ is connected;

b) $\Phi(0,y) = y$;

c) if (t',y), $(t+t',y)$, and $(t, \Phi(t',y))$ are in W then $\Phi(t+t',y) = \Phi(t, \Phi(t',y))$.

1.7. For $W = \mathbb{R}\times M$, Φ is a (global) <u>one-parameter group of diffeomorphisms</u> of M. In this case the map $\varphi_t : x \longmapsto \Phi(t,x)$ is a diffeomorphism of M for every $t\in\mathbb{R}$, and we have:

i) φ_o = identity map;

ii) $\varphi_{t+t'} = \varphi_t \circ \varphi_{t'}$;

iii) $(\varphi_t)^{-1} = \varphi_{-t}$.

In other words, Φ is a differentiable (analytic) action of $\mathbb{R}$ on M.

Such a flow will often be denoted by $(\varphi_t)_{t\in\mathbb{R}}$.

1.8. <u>Remark</u>. A vector field X on M allows the construction of a flow (W,Φ) of diffeomorphisms of M (of the same differentiability class as X) such that for every point y of M the curve $t \longmapsto \Phi(t,y)$ is an integral curve of X. (W,Φ) is a local one-parameter group <u>generated</u> by X.

The germ of such a flow along $\{0\}\times M$ is uniquely determined by X (corollary 1.5).

Conversely, if (W,Φ) is a flow of class C^r (respectively analytic) on M, there exists one and only one generating vector field X of class C^{r-1} (respectively analytic): the value of X at a point $y\in M$ is the vector tangent in y to the curve $t \longmapsto \Phi(t,y)$. This field whose value at the point y is $\frac{\partial\Phi}{\partial t}(0,y)$ is of class C^{r-1}. It is easy to see, by using properties of flows, that its integral curves are the maps $t \longmapsto \Phi(t,y)$.

1.9. <u>PROPOSITION</u>. The set of all flows generated by the vector field X, ordered by inclusion, has a unique maximal element.

The "union" of all flows generated by X is indeed itself a flow generated by X.

If this local one-parameter group is a global group, X is called a <u>complete</u> vector field.

1.10. <u>Examples</u>.

i) The vector field $X = \sum_i x_i \frac{\partial}{\partial x_i}$ is complete on $\mathbb{R}^m$: it generates the flow of all homotheties of $\mathbb{R}^m$ (the diffeomorphism φ_t is the homothety of ratio e^t).

ii) The maximal flow on $\mathbb{R}$ generated by the field $X = x^2 \frac{\partial}{\partial x}$ is given by

$$W = \left\{(t,x) \in \mathbb{R}\times\mathbb{R} \mid 1-tx > 0\right\},$$

$$(t,x) = \frac{x}{1 - tx} \; .$$

1.11. <u>Remarks</u>.

i) In the case of the covering map of example iv) of 1.2, the two vector fields X and $\tilde{X}$ are simultaneously complete or not complete.

ii) If the vector field X is complete and generates a flow $(\varphi_t)_{t\in\mathbb{R}}$ on M, one has $\varphi_t^T \circ X \circ \varphi_{-t} = X$ for every $t \in \mathbb{R}$: every diffeomorphism φ_t transforms any integral curve of X into another one.

It follows that for a submanifold N of codimension 1 of M which is transverse*) to X the submanifold $\varphi_t(N)$ remains transverse to X for every $t \in R$. Hence the map $(t,x) \longrightarrow \varphi_t(x)$ is then a submersion of $\mathbb{R} \times N$ into M.

---

* A submanifold N of codimension 1 is <u>transverse</u> to X if for every point y of N the vector X(y) does not belong to the tangent hyperplane $T_y(N)$ to N at y.

If dimension M = 2, and if N is diffeomorphic to the circle S^1 (or an interval of $\mathbb{R}$) one calls N a <u>closed transversal</u> (respectively a <u>transverse arc</u>) to X.

iii) Let Y be a complete vector field on the product $M \times \mathbb{R}$ of the form $Z(y,s) + \frac{\partial}{\partial s}$, $Z(y,s) \in T_y M$. Then the flow generated on $M \times \mathbb{R}$ may be written as $(t,y,s) \longmapsto (f_{t,s}(y), t+s)$, where $f_{t,s}$ is a family of diffeomorphisms of M satisfying the following properties:

a) $f_{0,s}$ = identity map for every s;

b) $f_{t+t',s} = f_{t,t'+s} \circ f_{t',s}$;

c) $(f_{t,s})^{-1} = f_{-t,t+s}$.

Letting $g_{t,s} = f_{t-s,s}$ we thus obtain a family of diffeomorphisms of M with the following properties:

a) $g_{s,s}$ = identity map for every s ;

b) $g_{t,s} \circ g_{s,r} = g_{t,r}$;

c) $(g_{t,s})^{-1} = g_{s,t}$.

This family is therefore determined by the "isotopy of the identity" $h_t = g_{t,0}$, because of $g_{t,s} = h_t \circ h_s^{-1}$.

Furthermore, keeping y and s fixed, the curve $t \mapsto g_{t,s}(y)$ is the solution of the non-autonomous system $z' = Z(z,t)$ with initial condition $z(s) = y$. The diffeomorphism $g_{t,s}$ may thus be interpreted as the translation of $M \times \{s\}$ to $M \times \{t\}$ along the integral curves of Y.

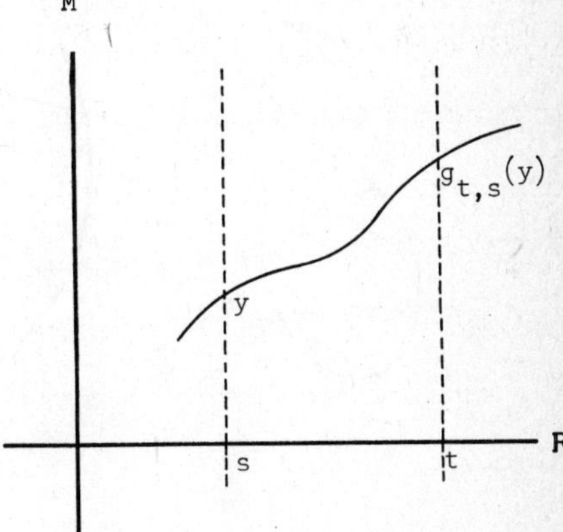

If Z is periodic with period τ we have $g_{t+\tau,s+\tau} = g_{t,s}$. Then the field Y induces a vector field on the cylinder $M \times S^1 = M \times (\mathbb{R}/\tau\mathbb{Z})$ (cf. exercise ii) of II-1.12).

1.12. <u>THEOREM</u>. A vector field with compact support is complete. (The <u>support</u> of a vector field is the closure of the set of its regular points.)

In particular every vector field on a compact manifold is complete.

Theorem 1.12 is an immediate consequence of the following lemma:

1.13. <u>LEMMA</u>. Let (W,Φ), with $W = \bigcup_{y\in M} (\alpha_y, \omega_y) \times y$, be the maximal flow generated by X, and let the curve $\Phi([0,\omega_x) \times x)$ be relatively compact for a certain point x of M. Then $\omega_x = +\infty$.

<u>Proof</u>.

Assume the curve $\Phi([0,\omega_x)\times\{x\})$ to be relatively compact and ω_x finite, and let y be a point of accumulation of $\Phi(t,x)$ for t tending towards ω_x.

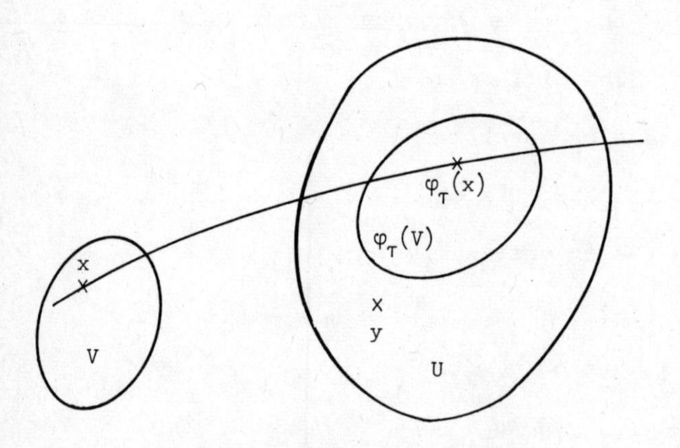

We choose an open neighbourhood U of y, a number ε between 0 and ω_x, and a differentiable map $\Psi: (-\varepsilon,+\varepsilon)\times U \to M$ having the properties a), b), and c) of theorem 1.4.

Let $\tau\in(\omega_x-\varepsilon, \omega_x)$ such that $\Phi(\tau,x) \in U$, and let V be an open neighbourhood of x such that $\{\tau\} \times V$ is contained in W, and $\Phi(\{\tau\} \times V)$ in U.

Under these conditions Φ can be extended to the open set $W \cup ((\omega_x - \varepsilon, \tau + \varepsilon) \times V)$ by defining $\Phi(t,z) = \Psi(t-\tau, \Phi(\tau,z))$ for $z \in V$ and $t \in (\omega_x - \varepsilon, \tau + \varepsilon)$.

We thus obtain a new flow generated by X and strictly greater than (W, Φ) which is contrary to the assumption of maximality. Q.E.D.

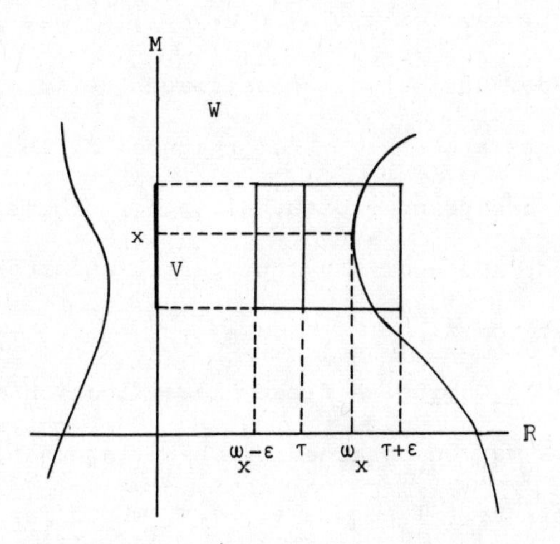

Still denoting by (W, Φ) the maximal flow generated by X we can draw the following conclusions from lemma 1.13:

1.14 <u>PROPOSITION</u>. Every integral curve of X assuming the value y for $t = \tau$ is the restriction of the map $t \longmapsto \Phi(t-\tau, y)$ to a subinterval of $(\alpha_y - \tau, \omega_y - \tau)$.

1.15 <u>COROLLARY</u>. A vector field X is complete if and only if every integral curve of X can be extended to an integral curve defined on all of $\mathbb{R}$.

1.16 PROPOSITION. Let Y be a vector field on the product $M \times \mathbb{R}$ of the form $Z(y,s) + \frac{\partial}{\partial s}$, with $Z(y,s) \in T_yM$, and let the projection onto M of the support of Z be relatively compact. Then the field Y is complete.

1.17. In view of proposition 1.14 the curve c_y defined on the interval (α_y, ω_y) by $t \mapsto \Phi(t,y)$ is called the maximal integral curve of X passing through y.

The images of any two maximal integral curves are either disjoint or they coincide. The set of these images defines thus a partition of M; its elements are called the orbits of X. In particular the singular points of X are point orbits. All other orbits are called non-singular. The quotient space of M by the partition of orbits of X is called the orbit space of X.

1.18. Classification of Orbits. A first classification of the orbits of the vector field X may be obtained by observing that, for a given orbit γ, all maximal integral curves c_y, $y \in \gamma$, which parametrize γ, satisfy simultaneously one of the following three properties:

i) c_y is injective;

ii) c_y is neither injective nor constant;

iii) c_y is constant.

In the first case the map c_y is an injective immersion of the interval (α_y, ω_y) into M; but it is not necessarily an embedding: there exist, e.g. on the torus T^2, vector fields all of whose orbits are everywhere dense (cf. example 4.12).

In the second case, if $c_y(b) = c_y(a)$, $b > a$, the map c_y is defined on $\mathbb{R}$ (lemma 1.13), and is periodic with a minimal period τ which is a fraction of b-a (this period is evidently independent of the choice of the point y on γ). γ is then called a <u>periodic orbit</u> of X, of period τ; it is a sub-manifold of M diffeomorphic to the circle S^1.

Finally, in the third case, y is a singular point of X.

1.19. <u>PROPOSITION</u>. Let X be a vector field on the paracompact manifold M. There exists a strictly positive function f on M, of the same differentiability class as X, such that the vector field $Y = fX$ is complete.

<u>Proof</u>. Paracompactness of M implies the existence of a proper function g of class C^s on M. Let $f = \exp(-(Xg)^2)$. If $Y = fX$, then $|Yg| = |(Xg)\exp(-(Xg)^2)| \leq 1$ on M. If c denotes an integral curve of Y defined on a bounded interval J, then $\frac{d}{dt}(g \circ c) = (Yg) \circ c$; hence $|\frac{d}{dt}(g \circ c)| \leq 1$ on J.

Thus the image of $g \circ c$ is bounded and hence the image of c is relatively compact. The proof is concluded by an application of lemma 1.13.

Q.E.D.

1.20. <u>Remark</u>. If c is the maximal integral curve of the vector field X passing through z, and if f is a never vanishing function on M, then the maximal integral curve of $Y = fX$ through z is the map $t \mapsto c(h(t))$, where h is the maximal solution of the differential equation $\frac{ds}{dt} = f(c(s))$ satisfying $h(0) = 0$. Thus these maximal integral curves differ only by a change of parameter, which preserves the orientation for positive f. Hence the orbits of X and Y coincide. We may thus assume, in what follows, the field to be complete (as far as properties of the orbits of a vector field on a paracompact manifold which are invariant with respect to parameter transformations are concerned).

As an example we have

1.21. PROPOSITION. The equivalence relation on M whose classes are the orbits of a vector field is open.

Indeed, denoting by $(\varphi_t)_{t\in R}$ the one-parameter group of diffeomorphisms of M generated by the given complete vector field, and by U an open set of M, then $\bigcup_{t\in \mathbb{R}} \varphi_t(U)$ (the "saturated set" of U) is open.

1.22. Remark. Some of the preceding results may be extended to the case where M is a manifold with boundary, provided that the vector field X has a "sufficiently nice" behaviour on the boundary of M; for example if it is either tangent or transverse to the boundary.

In particular, if X is a vector field on M transverse to the boundary and pointing inward along the boundary, one obtains, by restriction to positive times, a local one-parameter semigroup of diffeomorphisms of M.

Using a partition of unity in the case of a paracompact manifold M one can construct such an inward pointing vector field which is "positively complete". Hence:

1.23. THEOREM. Let M be a paracompact manifold with boundary. There exists a diffeomorphism h of $\partial M \times [0,+\infty)$ on an open set V of M satisfying $h(y,0) = y$ for all $y \in \partial M$.

V is then called a collar of the boundary of M.

1.24. Exercise. Let M and N be two compact manifolds with boundary of the same dimension, and let h be a diffeomorphism of a component of the boundary of M onto a component of the boundary of N. Then there exists on the adjunction space $V = M \cup_h N$ (obtained by glueing M to N via h) a structure of differentiable manifold, unique up to a diffeomorphism, for which the canonical injections of M and N into V are embeddings.

If in particular $M = N$, and h the identity map of ∂M, then the manifold V is called the double of M.

There is a similar result obtained by identifying through a diffeomorphism two connected components of the boundary of a manifold M.

2. GENERAL THEORY OF ORBITS

The fundamental tool for the study of non-singular orbits of a vector field is the following theorem (flow box theorem):

2.1. THEOREM. Let X be a vector field of class C^r (respectively analytic) on M and let y be a regular point of X. Then there exists in a neighbourhood of y a local coordinate system $(y_1,\ldots,y_m)$ of class C^r (respectively analytic) in which X is given by the expression $\frac{\partial}{\partial y_1}$.

(Such a local system of coordinates, together with a corresponding open set $a_i < y_i < b_i$, $i=1,\ldots,m$, is called distinguished for X.)

Proof. Since the problem is local we may restrict ourselves to the case where X is a vector field on $\mathbb{R}^m$ satisfying $X(0) = \frac{\partial}{\partial x_1}$. Now we denote by (W,Φ) a local one-parameter group generated by X, and by h the map defined in a neighbourhood of 0 in R^m by $h(x_1,x_2,\ldots,x_m) = \Phi(x_1,0,x_2,\ldots,x_m)$.

The Jacobian matrix of h in 0 is the identity matrix. Hence the map h has in a neighbourhood of 0 an inverse which defines a local system of coordinates $(y_1,\ldots,y_m)$ in which X is given by $\frac{\partial}{\partial y_1}$ because of $h^T(\frac{\partial}{\partial x_1}) = X \circ h$. Q.E.D.

2.2. Remarks.

i) Let $(y_1, \ldots, y_m)$ and $(z_1, \ldots, z_m)$ be two distinguished local systems of coordinates for X. Then the coordinate transformation $z_i = f_i(y_1, \ldots, y_m)$, $i = 1, 2, \ldots, m$, satisfies the conditions $\frac{\partial f_1}{\partial y_1} = 1$ and $\frac{\partial f_i}{\partial y_1} = 0$ for $i \geq 2$.

ii) Since every non-singular orbit of X is the image of the line $\mathbb{R}$ or of the circle S^1 under an injective immersion the intersection of such an orbit and of a distinguished open set U is a countable union of slices y_2 = constant, y_3 = constant, ... , y_m = constant in U.

2.3. DEFINITION. A non-singular orbit γ of X is proper (respectively locally dense) if there is a distinguished open set U for X such that the intersection $U \cap \gamma$ is a single slice of U (respectively is dense in U). A recurrent orbit is a non-singular orbit which is not proper.

These concepts are obviously invariant under coordinate transformations.

2.4. PROPOSITION. Let γ be a proper (respectively locally dense) orbit of X. Then every point y of γ has an open distinguished neighbourhood V for X such that $V \cap \gamma$ is a single slice of V (respectively is dense in V).

Indeed, denoting by (W, Φ) the maximal flow generated by X, we can choose the open set U in definition 2.3 and a point $z \in U \cap \gamma$ in such a way that $\{t\} \times U \subset W$ for $y = \Phi(t,z)$. The open set $V = \Phi(\{t\} \times U)$ has the desired property.

2.5. <u>COROLLARY</u>. A non-singular orbit is locally dense if and only if there is an open set of M in which it is dense.

2.6. <u>PROPOSITION</u>. A non-singular orbit γ of X is proper if and only if it is a locally closed submanifold of M.

<u>Proof</u>. The necessity of the condition follows from proposition 2.4. Assume now that γ is a locally closed submanifold of M, and let V be a distinguished open set for X such that $V \cap \gamma$ is closed in V. Then one of the slices of $V \cap \gamma$ is isolated in the set of slices of $V \cap \gamma$, because a countable closed set of $\mathbb{R}^{m-1}$ has an isolated point (Baire property). We can thus construct a distinguished open set U in V containing only this particular slice of $V \cap \gamma$. Q.E.D.

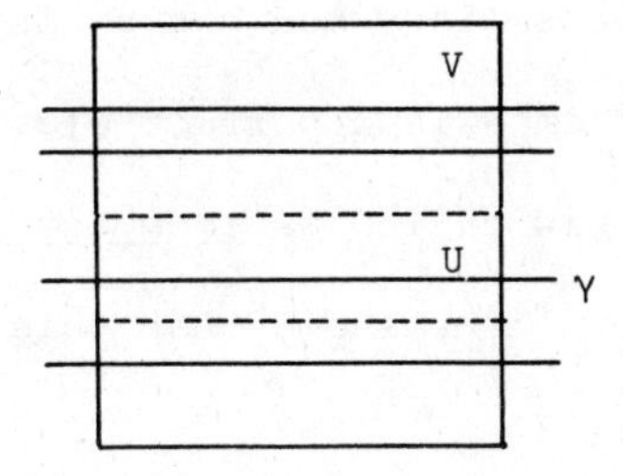

Consequences:

i) A closed non-singular orbit (in particular a periodic orbit) is proper;

ii) a non-singular and non-periodic orbit is proper if and only if each of its parametrizing maximal integral curves is an embedding (which is not necessarily proper) of $\mathbb{R}$ into M;

iii) a non-singular and compact orbit is periodic.

2.7. <u>Remark</u>. Let q: $\tilde{M} \to M$ be a covering map of M, and let $\tilde{X}$ be the vector field on M determined by X (cf. example iv) of 1.2), $\tilde{\gamma}$ an orbit of $\tilde{X}$ and $\gamma = q \circ \tilde{\gamma}$ the correponding orbit of X.

Then the following statements hold:

i) If $\tilde{\gamma}$ is periodic or locally dense, then so is γ;

ii) if γ is proper, so is $\tilde{\gamma}$.

Furthermore, if $\tilde{M}$ is a finite covering of M the converse statements are also true.

2.8. The case of surfaces. If M is a surface (i.e. a manifold of dimension 2), and if N is an open arc transverse to X, then a non-empty intersection I of N and of a non-singular orbit γ of X is a countable set having one of the following three properties:

i) I is discrete in N, in which case γ is proper;

ii) I is dense in an open set of N; then γ is locally dense;

iii) I is nowhere dense in N and has no isolated point. In this case the closure $\bar{I}$ of I in N is a perfect closed set (homeomorphic to a Cantor space). Then γ is called an exceptional orbit.

In chapter IV it will be shown that this case can occur on a torus T^2 (example ii) of 4.12).

We finish this section by two technical results which will be useful later on.

2.9. PROPOSITION. Let c be an integral curve of X which is defined on a compact interval $[a,b]$, and let c be an injective map. Then there exists a distinguished open set for X which contains the image of c.

Proof. Denote by (W, Φ) the maximal flow generated by X. We can determine a number $\epsilon > 0$ and an embedding f of $\mathbb{R}^{m-1}$ into M satisfying the following properties:

i) $f(\mathbb{R}^{m-1})$ is transverse to X;

ii) $f(0) = c(a)$;

iii) $(-\varepsilon, b-a+\varepsilon) \times f(\mathbb{R}^{m-1}) \subset W$.

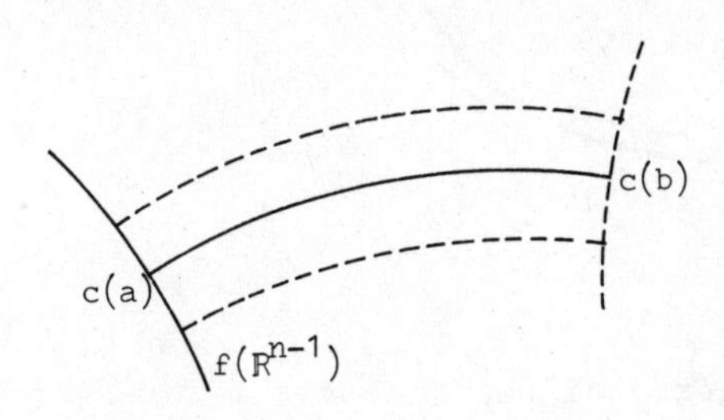

Denote now by h the map of $(-\varepsilon, b-a+\varepsilon) \times \mathbb{R}^{m-1}$ into M defined by $h(t,x) = \Phi(t,f(x))$. This map is a submersion, and we have

$$h(t,0) = c(t+a) \quad \text{for} \quad t \in [0,b-a].$$

Hence there is an open neighbourhood V of $[0,b-a] \times 0$ in $\mathbb{R} \times \mathbb{R}^{m-1}$ such that the restriction of h to V is a diffeomorphism of V onto a distinguished open set for X containing the image of c. Q.E.D.

2.10. PROPOSITION. Let X be a nowhere vanishing vector field on an orientable surface M. Then every non-closed orbit of X is intersected by a closed transversal to X.

Proof. Let γ be a non-closed orbit of X, x a point of $\overline{\gamma} - \gamma$, and U a distinguished open neighbourhood of x for X. We may then choose points y and z in the given order of the oriented orbit γ, belonging to distinct slices of $U \cap \gamma$ and in such a way that the arc of γ with initial point y and terminal point z does not intersect any other slice of $U \cap \gamma$.

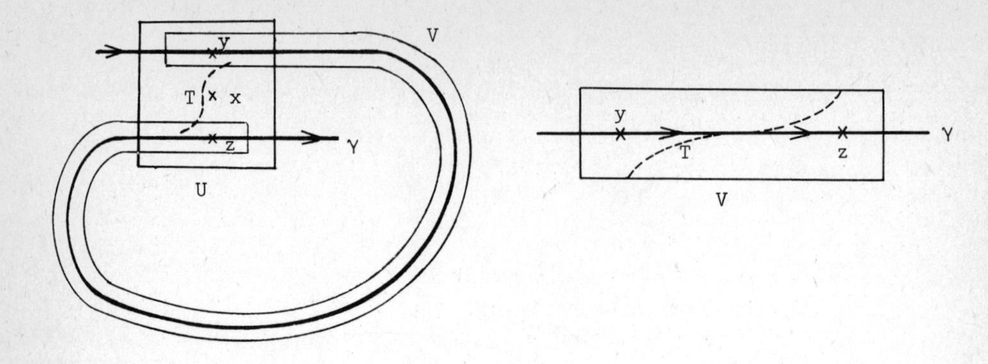

Having chosen a distinguished open set V for X such that y and z are in the same slice of $V \cap \gamma$ (cf. proposition 2.9) the above figures show how to construct a closed transversal T to X intersecting γ. The assumption of orientability of M implies that V is glued to U as in a cylinder and not as in a Moebius strip. Q.E.D.

2.11. Remarks.

i) The saturated set of T for the orbits of X coincides with the saturation of the "segment" yz of U.

ii) In a similar way we can construct a closed transversal intersecting a non-singular orbit which has two distinct slices in a distinguished open set, or which cuts a transverse arc in two distinct points.

3. INVARIANT AND MINIMAL SETS.

3.1. Assumption. In the rest of this chapter we shall be dealing with properties of orbits of a vector field X. We will assume that X is complete which is no restriction for a paracompact M, in view of remark 1.20. We then denote by $(\varphi_t)_{t \in R}$ the flow generated by X.

3.2. A subspace A of M is a union of orbits of X if and only if $\varphi_t(A)=A$ for every $t \in \mathbb{R}$. For this reason A is called invariant with respect to the vector field X (or with respect to the flow φ_t).

(More generally a geometric object on M (e.g. a differential form, a sub-fiber bundle of $\Gamma(M)$ etc.) is called invariant with respect to X if it is invariant with respect to the flow φ_t.)

The complement, the closure, the interior, and the boundary of an invariant subspace A are invariant as well.

In particular, the closure $\bar{\gamma}$ of an orbit γ is an invariant closed and connected set. An orbit contained in $\bar{\gamma}$ is then called adherent to γ.

3.3 Exercises.

i) A non-singular orbit γ is proper if and only if γ is an open set in $\bar{\gamma}$.

ii) If γ is a locally dense orbit adherent to an orbit γ', then γ' is also locally dense, and $\overline{\gamma'} = \bar{\gamma}$.

iii) If M is a connected manifold, and if all orbits of X are locally dense, then they are all everywhere dense.

3.4. DEFINITION. A minimal set for X is a minimal element of the set of all those subspaces of M, ordered by inclusion, which are invariant, closed, and non-empty.

Hence the following statements hold:

i) Every orbit which is contained in a minimal set is dense in it;

ii) every minimal set is connected;

iii) two distinct minimal sets are disjoint.

3.5. Example. A singular point or a closed orbit, in particular a periodic orbit, is a minimal set.

3.6. Exercises.

i) A minimal set on a connected manifold M different from M does not have interior points.

ii) If M is a connected surface a minimal set E is of one of the following four types:

- a singular point;
- a closed non-singular orbit (periodic if M is compact);
- the whole manifold M (in this case all orbits of X are everywhere dense);
- the closure of an exceptional orbit; in this case all orbits contained in E are exceptional. E is then called an exceptional minimal set.

3.7. THEOREM. Every invariant compact and non-empty set contains a minimal set.

(On the other hand M. Herman has constructed on $T^2 \times \mathbb{R}$ an analytic vector field without a minimal set.)

Indeed let A be such a subspace; then the set of all closed invariant and non-empty subspaces contained in A is inductive (the intersection of a nested family of compact and non-empty sets is not empty). By Zorn's lemma it has therefore a minimal element.

3.8. COROLLARY. If the manifold M is compact, and if all non-singular orbits of X are proper of locally dense, then the closure of every proper orbit contains a singular point or a periodic orbit.

Proof. Let γ denote a proper orbit of X. Its closure $\bar{\gamma}$ contains a minimal set E. If E is not a singular point of X then all orbits contained in E are proper (cf. exercis ii) of § 3.3); hence E reduces to a periodic orbit (exercise i) of § 3.3). Q.E.D.

3.9. Exercise. Construct (e.g. on the torus T^3) a vector field having a proper orbit whose closure contains non-proper orbits.

3.10. Remark. More generally than above one can introduce the concepts of invariant subspaces and minimal sets for an arbitrary equivalence relation on a topological space. Theorem 3.6 then remains valid.

Furthermore for an open equivalence relation (e.g. the relation associated to a group action, or a foliation) the closure of an invariant subspace A is itself invariant: indeed, if $\bar{A}$ were not invariant the saturated set of its complement would be an open set intersecting A.

4. LIMIT SETS.

For every point y of M we denote

$\gamma_y = \{\varphi_t(y),\ t \in \mathbb{R}\}$ = the orbit of y,

$\gamma_y^+ = \{\varphi_t(y),\ t \geq 0\}$ = the positive half-orbit of y,

$\gamma_y^- = \{\varphi_t(y),\ t \leq 0\}$ = the negative half-orbit of y.

These concepts are obviously invariant with respect to parameter transformations preserving the orientation of the orbits.

4.1. DEFINITION. Let γ be an orbit of X. The set $\Omega_\gamma = \bigcap_{y \in \gamma} \overline{\gamma_y^+}$ (respectively $A_\gamma = \bigcap_{y\in\gamma} \overline{\gamma_y^-}$) is called the <u>ω-limit set</u> of γ (respectively its <u>α-limit set</u>).

4.2. <u>Remarks</u>.

i) Interchanging t and -t (i.e. X and -X) leads to an interchange of ω- and α-limits.

ii) The set Ω_γ (respectively A_γ) consists of the limit points of the maps $t \mapsto \varphi_t(y)$, $y \in \gamma$, for t tending towards $+\infty$ (respectively $-\infty$). Hence we have:

4.3. <u>PROPOSITION</u>. The limit sets are closed and invariant.

4.4. <u>Examples</u>.

i) For γ a singular point or a periodic orbit we have $\Omega_\gamma = A_\gamma = \gamma$.

ii) If γ is a non-singular, closed, and non-periodic orbit we have $\Omega_\gamma = A_\gamma = \emptyset$.

iii) If M is a connected manifold, and if γ is an everywhere dense orbit then $\Omega_\gamma = M$ or $A_\gamma = M$.

iv) For a non-singular and non periodic orbit γ the following properties are equivalent:

γ is proper;

$\gamma \cap A_\gamma = \gamma \cap \Omega_\gamma = \emptyset$;

$\bar{\gamma} - \gamma = A_\gamma \cup \Omega_\gamma$.

v) For a non-singular orbit γ the following properties are equivalent:

γ is recurrent (i.e. not proper);

$\gamma \subset \Omega_\gamma$ or $\gamma \subset A_\gamma$;

$$\bar{\gamma} = \Omega_\gamma \cup A_\gamma \ .$$

vi) For a non-singular and non-periodic orbit γ we have $\Omega_\gamma \cup A_\gamma = \overline{\bar{\gamma} - \gamma}$.

4.5 Exercise: Gradient Field.

Let g denote a Riemannian metric on M, and f a differentiable function defined on M. Then the gradient of f is the vector field on M, denoted by grad f, which is characterized by the following property:

For any vector field Y on M we have $g(Y, \text{grad } f) = Yf$.

Under these conditions we have:

i) The singular points of grad f are the critical points of f (i.e. the points with $df = 0$);

ii) The function f is strictly increasing on every non-singular orbit of grad f; hence this vector field has no periodic orbits;

iii) The function f is constant on the limit sets of the non-singular orbits of grad f; hence these limit sets consist of critical points of f.

4.6. DEFINITION. An orbit γ of X is ω-stable (respectively α-stable) in the sense of Lagrange if there is a point y of γ such that the half-orbit γ_y^+ (respectively γ_y^-) is relatively compact.

If this is so then the same property holds for all points of γ.

4.7. PROPOSITION. If γ is an ω-stable (respectively α-stable) orbit in Lagrange's sense then the limit set Ω_γ (respectively A_γ) is compact, connected, and non-empty.

Proof. Taking into account the remark i) of §4.2 we can restrict ourselves to the case where γ is ω-stable in Lagrange's sense. Then every half-orbit γ_y^+, $y \in \gamma$, is relatively compact, and hence $\Omega_\gamma = \bigcap_{y\in\gamma} \overline{\gamma_y^+}$ is non-empty and compact. If Ω_γ were not connected we could find two non-empty

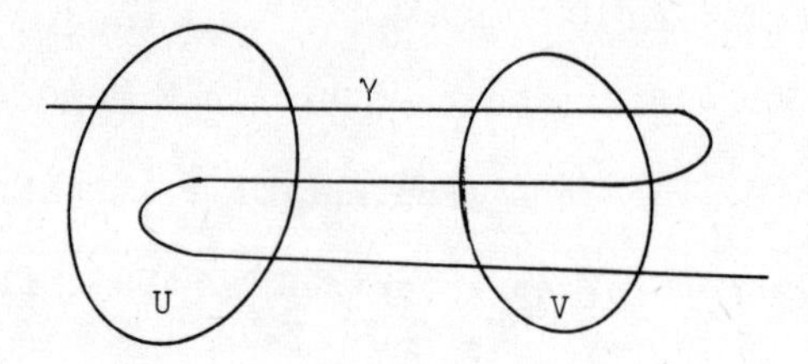

disjoint open sets U and V covering it, and for a given point y of γ, a sequence (t_n) tending towards $+\infty$ such that the sequence $\varphi_{t_n}(y)$ belongs to $M - (U \cup V)$ and is convergent, because of ω-stability. But then this limit would not be in Ω_γ, which is absurd. Q.E.D.

4.8. Remark. The example sketched below shows that the conclusion of 4.7 concerning the connectedness of Ω_γ need not hold if γ is not ω-stable in Lagrange's sense.

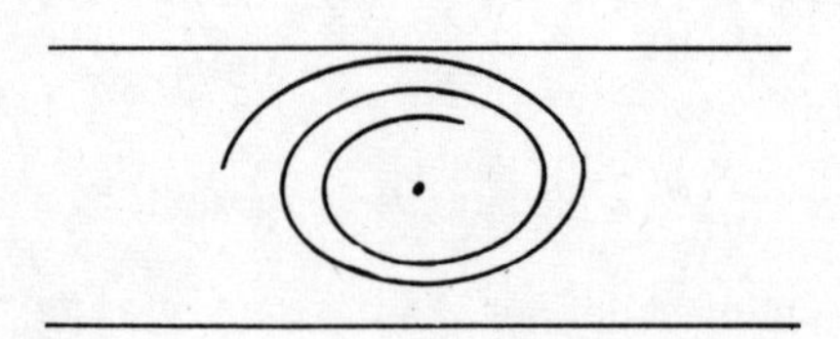

4.9. A given diffeomorphism f of a manifold N leads to a discrete one-parameter group of diffeomorphisms of N (i.e. a differentiable action of $\mathbb{Z}$ on N) by letting $(u,x) \longmapsto f^n(x)$.

The concepts and results of the preceding paragraphs extend to such an action (with the exception of those concerning connectedness). The following construction actually provides a connection between discrete and continuous one-parameter groups.

4.10. <u>Suspension of a diffeomorphism</u>. Let f be a diffeomorphism of a manifold N. The maps $(x,s) \longmapsto (f^{-n}(x), s+n)$, $n \in \mathbb{Z}$, define a proper and free action of $\mathbb{Z}$ on the product $N \times \mathbb{R}$. Let M denote the quotient manifold of $N \times \mathbb{R}$ by this action, and let N be identified with the image submanifold of $N \times \{0\}$ in M.

The vector field $\frac{\partial}{\partial s}$ on $N \times \mathbb{R}$ is invariant under this action. It determines a complete and nowhere vanishing vector field X on M, having the following properties:

i) The submanifold N is transverse to X;

ii) every orbit of X intersects N;

iii) for every point y of N we have $\varphi_1(y) = f(y)$.

Hence there is a one-to-one correspondence between the orbits of X in M and those of f in N. (The periodic orbits of X correspond to the periodic points of f, etc..)

M is called the <u>suspension manifold</u> of the diffeomorphism f, and X the <u>suspension field</u> of f. The field X is of class C^{r-1} if f is of class C^r, $r \geqslant 1$; but it generates a one-parameter group of class C^r.

4.11. <u>Exercises</u>.

i) The suspension of f is diffeomorphic to the manifold constructed from the product $N \times [0,1]$ by identification of the points $(x,1)$ and $(f(x),0)$ of its boundary (cf. exercise 1.24).

In particular, the suspensions of f and f^{-1} are diffeomorphic.

ii) The fundamental group of M is isomorphic to the extension of the group $\Pi_1(N)$ by $\mathbb{Z}$ corresponding to the automorphism f_* of $\Pi_1(N)$.

The cohomology space $H^1(M,\mathbb{R})$ is of dimension 1 if and only if 1 is not an eigenvalue of the homomorphism $f^*: H^1(N,\mathbb{R}) \longrightarrow H^1(N,\mathbb{R})$.

iii) Let g be a diffeomorphism of a manifold P, and let h denote a differentiable map of N into P satisfying $g \circ h = h \circ f$. The map $(x,s) \longmapsto (h(x),s)$ induces a smooth map k of the suspension of f in the suspension of g such that $k^T \circ X = Y \circ k$, where Y is the suspension field of g.

In particular for a diffeomorphism k (i.e. for f and g being conjugate) the suspensions of f and g are diffeomorphic.

iv) Let (h_s) be an isotopy of f to a diffeomorphism g of N (cf. definicion A.1); and let k be the map of $N \times \mathbb{R}$ into N coinciding with $g^{-n} \circ h^{-1}_{s-n} \circ f^n$ on $N \times [n,n+1]$, $n \in \mathbb{Z}$. The map $(x,s) \longmapsto (k(x,s),s)$ then induces a diffeomorphism K of the suspension of f on the suspension of g.

In particular for f isotopic to the identity map its suspension is isomorphic to the product $N \times S^1$.

4.12. <u>Examples</u>. If N is the circle S^1 the manifold M is diffeomorphic to the torus T^2 if the diffeomorphism f preserves orientation, and to the Klein bottle K^2 otherwise (a diffeomorphism of the circle is indeed isotopic to the identity if it preserves orientation, and to the symmetry $\theta \longrightarrow -\theta$ if it reverses orientation). In the latter case f has two fixed points; hence its suspension field has two periodic orbits of period 1.

If f is a rotation through the angle α of the circle $S^1 = \mathbb{R}/\mathbb{Z}$ the two following cases may be distinguished:

i) For a rational α all points of S^1 are periodic for f (of period q for $\alpha = p/q$); and all orbits of X are periodic;

ii) for an irrational α all orbits of f are dense in S^1, and the same holds for the orbits of X in T^2.

5. DIRECTION FIELDS.

5.1. A subbundle E of rank 1 of the tangent bundle T(M) is determined by giving a straight line E_y in the tangent space T_yM for each point $y \in M$. Such a subbundle is therefore called a direction field on M. Moreover this field is called orientable if the subbundle E is trivial.

An integral curve of a direction field E is an immersion h of an interval J of $\mathbb{R}$ into M such that $h'(t) \in E_{h(t)}$ for every $t \in J$.

5.2. Examples.

i) A vector field X on M determines an orientable field of directions on the open set U of its regular points: the fiber E_y of E above the point y of U is the subspace of $T_y(M)$ generated by the vector X(y). E is then called the field of the directions of the vector field X on U. The integral curves of E are, up to parameter changes, the integral curves of X.

Conversely let E be an orientable direction field on M. Then a nowhere vanishing vector field on M exists whose direction field coincides with E. Moreover for two such vector fields X and Y we have Y = fX, where f is a nowhere vanishing function on M. The set of all vector fields having E as direction fields can thus be partitioned into classes (numbering two for a connected M), where two elements of a class differ by a positive scalar factor. Each one of these classes is called an orientation of E, and E is called oriented if such a choice has been made.

ii) Let M be a parallelizable manifold. Choosing a trivialisation of T(M) a field of directions on M corresponds to a smooth map of M into the real projective space $\mathbb{P}\mathbb{R}^{m-1}$ of dimension m-1.

This direction field is orientable if and only if this map can be lifted to a map of M into the sphere S^{m-1}. (This property is of course independent of the choice of the parallelization of T(M).)

In particular every direction field on a simply connected open set of $\mathbb{R}^m$ is orientable.

iii) A Pfaffian form ω without singularities on a surface M determines a direction field E on M as follows: The fibre of E above a point y of M is the kernel of the linear form $\omega(y)$ in the tangent plane $T_y(M)$.

For an orientable and paracompact surface M this field E is orientable.

5.3. Exercise. The Pfaffian form $\omega_n = \sin \frac{n\theta}{2}\, d\rho + \rho \cos \frac{n\theta}{2}\, d\theta$, $\rho > 0$, $n \in N$, determines a direction field on the cylinder $R^2 - \{0\}$ which is orientable for n even, and non-orientable for n odd.

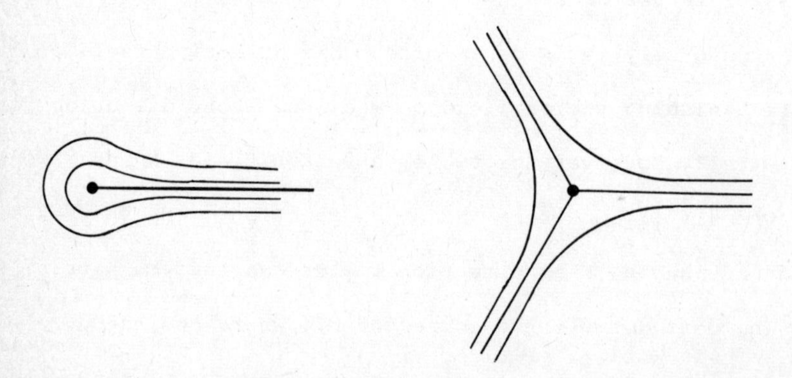

The integral curves are drawn above for the cases n=1 and n=3.

We will show now that every direction field on a paracompact manifold is orientable, up to a two-sheeted covering map.

5.4. <u>The covering space of the orientations of a direction field</u>.

Let E be a field of directions on the manifold M. Since E is orientable in a neighbourhood of every point of M there exists an atlas $(U_i, \varphi_i)_{i \in I}$ on M such that E is the field of the directions of the vector field $(\varphi_i^T)^{-1}(\frac{\partial}{\partial x_1})$ above U_i, and such that the parameter transformations $\varphi_j \circ \varphi_i^{-1} = (h_{ji}^1, \ldots, h_{ji}^m)$ satisfy $\frac{\partial h_{ji}^k}{\partial x_1} = 0$ for $k \geq 2$ (cf. theorem 2.1).

The maps $\gamma_{ji} = \frac{\partial h_{ji}^1}{\partial x_1} \circ \varphi_i \Big/ \left| \frac{\partial h_{ji}^1}{\partial x_1} \circ \varphi_i \right|$

define a cocycle on M with values in the group $\mathbb{Z}_2 = \{\pm 1\}$ [16]. The two-sheeted covering determined by this cocycle is denoted by $p: \hat{M} \longrightarrow M$, and $\hat{M}$ is called the <u>covering space of orientations</u> of E. (This covering space is, of course, independent of the choice of the atlas used above.)

5.5. <u>PROPOSITION</u>. The direction field $\hat{E} = p^*E$ (= inverse image of E under the projection p) is orientable.

In view of theorem V - 2.1 we conclude:

5.6. <u>COROLLARY</u>. The only compact and connected surfaces without boundaries on which there exists a direction field are the torus T^2 and the Klein bottle K^2.

<u>Proof of proposition 5.5</u>. Set S denote the topological sum (disjoint union) of the manifolds $U_i^\varepsilon = U_i \times \{\varepsilon\}$, $i \in I$ and $\varepsilon = \pm 1$. Then the manifold $\hat{M}$ is the quotient space of S by the equivalence relation identifying the points $(x, \varepsilon) \in U_i^\varepsilon$ and $(y, \eta) \in U_j^\eta$ for $x=y$ and $\eta = \gamma_{ji}(x)\varepsilon$.

Let s denote the symmetry $(x_1,x_2,\dots x_m)\mapsto(-x_1,x_2,\dots x_m)$ of $\mathbb{R}^m$. Then the maps

$$\Psi_i^{(+1)} = \varphi_i \circ p:\ U_i^{+1} \to R^m \text{ and}$$

$$\Psi_i^{(-1)} = s \circ \varphi_i \circ p:\ U_i^{-1} \to R^m$$

define an atlas on $\hat{M}$ for which the coordinate transformations $\Psi_j^{(\eta)} \circ (\Psi_i^{(\varepsilon)})^{-1} = (h^1_{ji\eta\varepsilon},\dots,h^m_{ji\eta\varepsilon})$ have the following properties:

i) $\dfrac{\partial h^k_{ji\eta\varepsilon}}{\partial x_1} = 0$ for $k \geqslant 2$;

ii) $\dfrac{\partial h^1_{ji\eta\varepsilon}}{\partial x_1} > 0$;

iii) $\dfrac{\partial h^1_{ji\eta\varepsilon}}{\partial x_1} \circ \Psi_i^\varepsilon$ is a cocycle defining the subbundle $\hat{E}$.

We now denote by X_i^ε the vector field on U_i^ε satisfying $(\Psi_i^\varepsilon)^T X_i^\varepsilon = \dfrac{\partial}{\partial x_1}$, and by (α_i^ε) a partition of unity which is subordinate to the open covering (U_i^ε); it is then easy to see that the vector field $\hat{X} = \sum \alpha_i^\varepsilon X_i^\varepsilon$ is a nowhere vanishing field on $\hat{M}$ which has $\hat{E}$ as its field of directions. Q.E.D.

5.7. <u>PROPOSITION</u>. The field of directions E is orientable if and only if its orientation covering is trivial.

The sufficiency is an immediate consequence of the preceding proposition. If E is orientable it is possible to choose in 5.4 an atlas on M for which the cocycle (γ_{ji}) admits only the value +1 (cf. remark i) of 2.2); hence follows the necessity.

5.8. <u>COROLLARY</u>. On a paracompact and simply connected manifold every direction field is orientable.

5.9. Orbits of a direction field. Let a direction field E on M be orientable. Then every integral curve of E is, up to a change of parameters, an integral curve of some vector field X on M having E as direction field.

If the field E is non-orientable then any integral curve of E is the projection on M of an integral curve of the direction field $\hat{E}$ onto the orientation covering $\hat{M}$ of E; hence for a paracompact M we are back in the preceding case.

We can thus define the orbits of the direction field E, viz. either as the orbits of a vector field X on M having E as field of its directions in the orientable case, or as the projections into M of the orbits of a vector field $\hat{X}$ on $\hat{M}$ having $\hat{E}$ as field of directions in the non-orientable case.

Most concepts and results of the preceding paragraphs then generalize to direction fields: e.g. periodic orbits, proper orbits, locally dense or exceptional orbits; invariant subspaces; minimal sets etc.

In particular such an investigation leads to the classification theorem of one-dimensional manifolds:

5.10. THEOREM. A connected and paracompact one-dimensional manifold is diffeomorphic to one of the following manifolds: the circle S^1, the line $\mathbb{R}$, the closed interval $[0,1]$, the half-open interval $[0,1)$.

Proof. Assume first M to have no boundary, and let $E = T(M)$ be the field of tangential directions to M.

Since every orbit of E is an open submanifold of M there is but a single one. If it is periodic then M is diffeomorphic to the circle S^1; if it is not periodic then M is diffeomorphic to the line $\mathbb{R}$.

If on the other hand M has now a non-empty boundary then its interior is diffeomorphic to $\mathbb{R}$. From theorem 1.23 we then conclude that a compact M is diffeomorphic to the interval $[0,1]$, and a non-compact M is diffeomorphic to $[0,1)$. Q.E.D.

5.11. Exercise. Let E be a direction field on a paracompact manifold M. The quotient bundle $F = T(M)/E$ is called the normal bundle to E. If this normal bundle is orientable then E is called transversally orientable (cf. [15])

i) For an orientable manifold M the field E is transversally orientable if and only if it is orientable. Conversely, if E is orientable and transversally orientable then the manifold M is orientable.

ii) If E is not transversally orientable then there exists a two-sheeted covering $q: \tilde{M} \to M$ of M such that the field of directions $\tilde{E} = q^*E$ (= inverse image of E under the projection q) is transversally orientable.

iii) For a surface M the following properties are equivalent:

E is transversally orientable;

F is trivial;

E can be defined by a Pfaffian form without singularity: F may be identified with a subbundle of rank 1 of the cotangent bundle $T^*(M)$.

iv) If E is a transversally orientable field of directions on a surface M every non-closed orbit of E is intersected by a closed transversal to E (cf. proposition 2.10).

In a similar context we also have the following result:

5.12. PROPOSITION. Every direction field on a compact surface M without boundary has a closed transversal.

Proof. Let F be a direction field on M which is complementary to the given field E (e.g. the field of orthogonal directions to E with respect to a Riemannian metric on M).

If F has no periodic orbit let γ be an oriented orbit (dense or exceptional) of a minimal set of F, and let V denote an open set which is distinguished simultaneously for E and for F and meets γ. As the figure below shows we then join two slices of γ in V which are "adjoining" and occur in the same sense, and thus obtain a closed transversal to E.

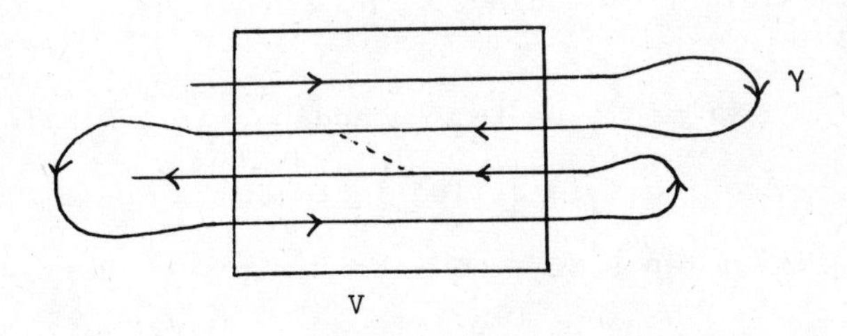

Q.E.D.

5.13. Exercise. Let M be a manifold without boundary, and denote by N a submanifold of codimension 1 of M. The normal bundle E to N in M is the quotient bundle of rank 1 on N of the bundle T(M) by the bundle T(N). Under these circumstances we have:

i) For E trivial there exists an open neighbourhood U of N in M which is diffeomorphic to $N \times \mathbb{R}$;

ii) if E is not trivial then there exists a two-sheeted non-trivial covering $\tilde{N}$ of N with canonical involution α, and an open neighbourhood U of N in M which is diffeomorphic to the quotient space of the product $\tilde{N} \times \mathbb{R}$ by the involution $(x,t) \longmapsto (\alpha(x),-t)$.

In these two cases U is called a <u>tubular neighbourhood</u> of N in M, and N is <u>two-sided</u> in the first situation, and <u>one-sided</u> in the second one.

In particular a simply connected N is always two-sided.

APPENDIX: VECTOR FIELDS AND ISOTOPIES.

Very often the analysis of a deformation problem can be reduced to the construction and integration of a "non-autonomous" vector field (cf. remark iii) of 1.11). We shall illustrate this point of view by studying isotopies of embeddings.

A.1. <u>DEFINITION</u>. Let f and g be two embeddings of a manifold N into a manifold M. The embedding f is called <u>isotopic</u> to the embedding g if there exists a smooth map F of $N \times \mathbb{R}$ into M with the following properties:

i) For every $t \in \mathbb{R}$ the map $f_t : x \longmapsto F(x,t)$ is an embedding of of N into M;

ii) we have $f_t = f$ for $t \leq 0$, and $f_t = g$ for $t \geq 1$.

The map $F = (f_t)$ is then called an <u>isotopy</u> from f to g.

A.2. <u>Exercises</u>.

i) Isotopy is an equivalence relation in the set of all embeddings of N into M.

ii) In the situation of exercise 1.24 the manifolds $M \cup_h N$ and $M \cup_k N$ are diffeomorphic if the diffeomorphism k is isotopic to h.

iii) A direct (i.e. orientation-preserving) embedding f of $\mathbb{R}^m$ into itself is isotopic to the identity: one can assume $f(0) = 0$, and consider the deformation of f defined by $x \mapsto \frac{1}{t} f(tx)$, $0 < t \leq 1$.

iv) Let M be a connected and oriented manifold of dimension m. Two direct embeddings f and g of $\mathbb{R}^m$ into M are isotopic: one can at first assume $f(\mathbb{R}^m) \cap g(\mathbb{R}^m) \neq \emptyset$.

v) Let M be a connected and oriented m-dimensional manifold without boundary. Two direct embeddings of the disk $\mathbb{D}^m$ into M are isotopic. If M is not orientable two arbitrary embeddings of $\mathbb{D}^m$ into M are isotopic.

vi) Let M and N be two compact, connected, and oriented manifolds of the same dimension m, and let f (respectively g) be a direct embedding of the disk $\mathbb{D}^m$ into the interior of M (respectively of N). The oriented manifold

$$\left[M - f(\mathbb{D}^m - S^{m-1})\right] \cup_{g \circ f^{-1}} \left[N - g(\mathbb{D}^m - S^{m-1})\right]$$

is (up to a diffeomorphism) independent of the choice of the embeddings f and g (cf. theorem A.5); it is called the <u>connected sum</u> of the manifolds M and N and is denoted by $M \# N$.

An analogous construction exists for M and N non-orientable.

This connected sum is commutative and associative up to diffeomorphisms, and it has the sphere S^m as a neutral element, again up to a diffeomorphism.

A.3. DEFINITION. An <u>isotopy of a manifold</u> M is a smooth map H of $M \times \mathbb{R}$ into M with the following properties:

i) For every $t \in \mathbb{R}$ the map h_t: $y \mapsto H(y,t)$ is a diffeomorphism of M;

ii) h_o is the identity map of M.

The <u>support</u> of the isotopy H is the closure of the complement of the set $I = \{y \in M \mid H(y,t) = y \ \forall\ t \in \mathbb{R}\}$ of the invariant points of M under H.

A.4. Remark. An isotopy of M is induced by a complete vector field on the product $M \times \mathbb{R}$ of the form $Z(y,s) + \frac{\partial}{\partial s}$, where $Z(y,s) \in T_y(M)$ (cf. remark iii) of 1.11).

A.5. THEOREM. Let N denote a compact manifold, and f and g two embeddings of N into a manifold M without boundary. If f and g are isotopic then there exists an isotopy with compact support $H = (h_t)$ of M satisfying $g = h_1 \circ f$.

This result allows us to consider a deformation of the embedding f of N into M as induced by a "deformation of the ambient manifold" M.

Proof. Let $F: N \times \mathbb{R} \to M$ be an isotopy from f to g, and denote by V a compact neighbourhood of $F(N \times \mathbb{R})$ in M.

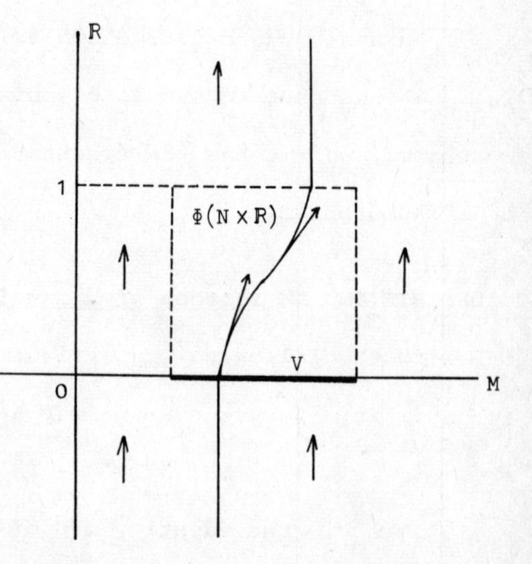

Since the map $\Phi: (x,s) \longmapsto (F(x,s),s)$ is an embedding of $N \times \mathbb{R}$ into $M \times \mathbb{R}$ we can construct a vector field X on $M \times \mathbb{R}$ with the following properties:

i) X is of the form

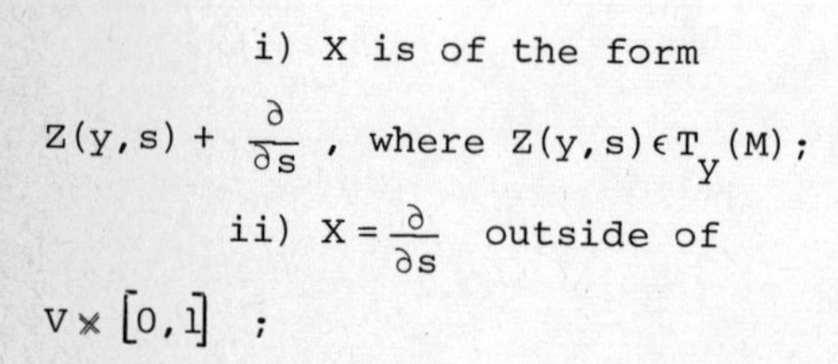

$Z(y,s) + \frac{\partial}{\partial s}$, where $Z(y,s) \in T_y(M)$;

ii) $X = \frac{\partial}{\partial s}$ outside of $V \times [0,1]$;

iii) $X(\Phi(x,s)) = \Phi^T(\frac{\partial}{\partial s})$.

(On the sketch above the field $Y = \frac{1}{4}X$ is drawn.)

The integration of this field then leads to the desired isotopy. Q.E.D.

A.6. <u>COROLLARY</u>. The group of diffeomorphisms of a connected manifold M without boundary acts transitively on M.

(More generally this group even acts transitively on the n-tuples of distinct points of M.)

A.7. <u>Exercises</u>. The following situations illustrate also the use of non-autonomous vector fields for deformation problems. (This method was introduced by J. Moser for solving the first problem: Trans. Amer. Math. Soc. 120, 1965, p. 286 - 294. It was subsequently applied to the following two problems by A. Weinstein and J. Martinet.)

The following preliminary result will be useful: denote by $(h_t)_{t\in\mathbb{R}}$ the isotopy of a manifold M generated by a "non-autonomous" vector field $Z_t(y) = Z(y,t)$ on M, and by $(\omega_t)_{t\in\mathbb{R}}$ a smooth one-parameter family of differential forms on M. Then

$$\frac{d}{dt}(h_t^*\omega_t) = h_t^*(\frac{d}{dt}\omega_t + L_{Z_t}\omega_t).$$

i) Let M denote a compact, connected, and oriented manifold, and ω_o and ω_1 two volume forms on M having equal integrals. Then there exists a diffeomorphism k of M which is isotopic to the identity map and satisfies $\omega_1 = k^*\omega_o$. (If α is a differential form of degree m-1 with $\omega_o - \omega_1 = d\alpha$ consider the family $\omega_t = (1-t)\omega_o + t\omega_1$ of volume forms, and define a non-autonomous vector field by $i_{Z_t}\omega_t = \alpha$.)

ii) Let $\omega_t = \omega_o + d\alpha_t$ denote a smooth one-parameter family of symplectic cohomologuous forms on a compact and connected manifold M of even dimension 2n (cf. [15]). Then there exists a diffeomorphism k of M which is isotopic to the identity map and satisfies $\omega_1 = k^*\omega_o$.

Analoguously, the <u>theorem of Darboux</u> may be proved, stating that a symplectic form may be expressed locally as $\sum_{i=1}^{n} dy_{2i-1} \wedge dy_{2i}$.

iii) Let $(\omega_t)_{t\in\mathbb{R}}$ denote a smooth one-parameter family of contact forms on a compact and connected manifold M (cf. [15]). Then there exists a diffeomorphism k of M which is isotopic to the identity map and such that the forms ω_1 and $k^*\omega_o$ define the same contact structure (i.e. the same field of hyperplanes on M): show that there is a uniquely defined non-autonomous vector field Z_t on M satisfying

$$\omega_t(Z) = 0 \quad \text{and} \quad \omega_t \wedge \left(\frac{d}{dt}\omega_t + L_{Z_t}\omega_t\right) = 0 .$$

Chapter II. The Local Behaviour of Vector Fields

In this chapter we shall present some properties of linear differential equations in $\mathbb{R}^m$ in order to study locally the orbits of a vector field X on a manifold M in a neighbourhood of a singular point or of a periodic orbit.

1. STABILITY AND CONJUGATION

1.1. DEFINITION. Let γ denote a singular point or a periodic orbit of X. Then γ is called ω-stable (respectively α-stable) in the sense of Liapounov if every neighbourhood U of γ contains a neighbourhood V of γ such that the positive (respectively negative) half-orbit of every point of V lies in U.

If it is possible, moreover, to choose this neighbourhood V in such a way that the ω-limit (respectively α-limit) set of each orbit meeting V is γ, then this stability is called asymptotic.

In a similar way stability properties in the sense of Liapounov are defined for a fixpoint or a periodic point of a local diffeomorphism f of a manifold N.

1.2. Remarks.

i) The passage from X to -X, or from f to f^{-1}, interchanges ω- and α-stability. We may therefore restrict ourselves to a study of the former property.

ii) A sufficiently close orbit to an ω-stable orbit in Liapounov's sense is ω-stable in the sense of Lagrange. The corresponding maximal integral curves are thus defined on the interval $[o,+\infty)$.

iii) Let γ be an asymptotically ω-stable orbit in Liapounov's sense. Then the set of all orbits having γ as ω-limit set is open in M, and is called the <u>basin</u> (of attraction) of γ.

iv) Under a finite covering map the stability properties in the sense of Liapounov are preserved (cf. example iv) of I-1.2).

v) Let X be the suspension field of a diffeomorphism f (cf. I-4.10). Then stability properties of periodic points of f and periodic associated orbits of X correspond to each other.

1.3. <u>Examples</u>.

i) The origin of $\mathbb{R}^2$ is ω- and α-stable in Liapounov's sense for the vector field with components $-y$ and x.

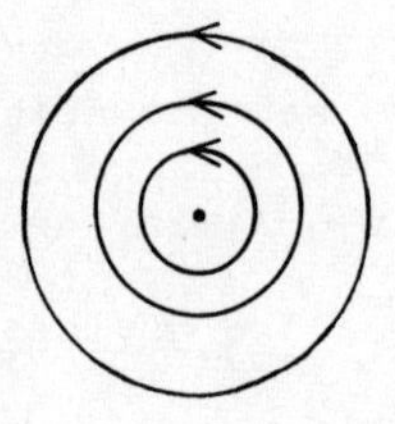

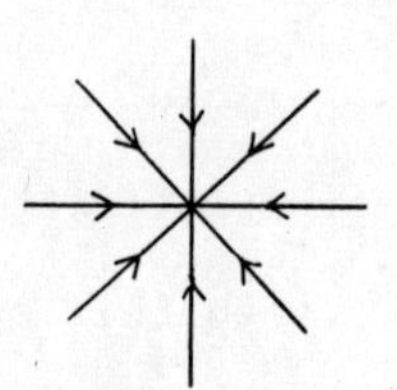

ii) The origin of $\mathbb{R}^m$ is asymptotically ω-stable in Liapounov's sense for the vector field $X: y \mapsto -y$.

iii) Let f be a diffeomorphism of $\mathbb{R}^m$ which is a <u>contraction</u> (i.e. satisfies $\|f(x)\| < \|x\|$ for every $x \neq 0$). Then the origin is a fixpoint which is asymptotically ω-stable in Liapounov's sense.

This holds in particular for a linear map f all of whose eigenvalues have absolute value less than 1.

1.4. DEFINITION. Let y be a singular point of the vector field X. A function L of class C^1 on an open neighbourhood U of y is called a <u>Liapounov function</u> for X in y if it satisfies the following properties:

i) $L(y) = 0$;

ii) $L > 0$ on $U - \{y\}$;

iii) $X.L \leqslant 0$ on U.

L is called a <u>strict Liapounov function</u> if, in addition, we have $X.L < 0$ on $U - \{y\}$.

The above condition iii) implies that for an integral curve c of X in U the function $L \circ c$ is decreasing.

1.5. <u>Example</u>. The square of the Euclidean norm is a Liapounov function (respectively a strict Liapounov function) at the origin for the vector field of example i) (respectively ii)) in 1.3.

1.6. <u>PROPOSITION</u>. If the field X has a Liapounov function L in a singular point y then this point is ω-stable in Liapounov's sense. Furthermore for a strict Liapounov function L this stability is asymptotic. (This characterization of the stability of a singular point is known as the <u>direct method of Liapounov</u>.)

<u>Proof</u>. The statement being a local one we may assume that M is the space $\mathbb{R}^m$, and that y is its origin.

Choose now $\varepsilon > 0$ sufficiently small such that L is defined on the closed ball B of centre 0 and radius ε, and let $\mu > 0$ denote the greatest lower bound of L on the boundary of B. Then η can be found with $0 < \eta < \varepsilon$ and $L(x) < \mu$ for $\|x\| \leqslant \eta$.

For an integral curve c of X satisfying $\|c(t_o)\| \leqslant \eta$ we have

$L(c(t)) < \mu$ for all $t \geq t_o$. Hence the image of c is contained in B, and the origin is ω-stable in Liapounov's sense.

Assume now L to be a strict Liapounov function, and let α denote a number between 0 and η, and ρ (respectively σ) the greatest lower bound of L (respectively of $|X.L|$) on the annulus $C = \{x \mid \alpha \leq \|x\| \leq \varepsilon\}$.

For sufficiently large t (cf. remark ii) of 1.2) we have $L(c(t)) \leq L(c(t_o)) - (t - t_o)\sigma < \rho$, hence $\lim_{c \to +\infty} c(t) = 0$. Q.E.D.

1.7. COROLLARY (Stability theorem of Lagrange). In a classical Hamiltonian system a strict local minimum of the potential is a stationary ω- and α-stable point in Liapounov's sense.

Proof. A classical Hamiltonian vector field X is locally a vector field on the space $\mathbb{R}^{2n}$ with coordinates $q_1, q_2, \ldots, q_n, p_1, p_2, \ldots, p_n$, and its components are of the form $\frac{\partial H}{\partial p_1}, \ldots, \frac{\partial H}{\partial p_n}, -\frac{\partial H}{\partial q_1}, \ldots, -\frac{\partial H}{\partial q_n}$, where H is a function on $\mathbb{R}^{2n}$ representable as a sum of two functions T and V satisfying the following properties:

i) T (the kinetic energy) is a positive definite quadratic form in $p_1, \ldots, p_n$, for $q_1, \ldots, q_n$ fixed;

ii) V (the potential) is independent of the variables $p_1, \ldots, p_n$.

The zeros of X are therefore the points (q,0), where q is a critical point of the potential V.

If, however, q is a strict local minimum of V (which may be assumed to vanish) then the Hamiltonian H is a Liapounov function for X at the singular point (q,0) because of $X.H = 0$. Q.E.D.

1.8. Exercise. Let X be the gradient field of a smooth function f (cf. exercise I-4.5). If the critical points of f are isolated, the only

ω-stable singular points of X in Liapounov's sense are the maxima of f, and they are asymptotically stable if they are isolated.

Furthermore, if the critical points of f are non-degenerate, the union of the basins of attraction of these maxima is an open and dense set on M.

1.9. DEFINITION. A vector field X on M is called C^t-conjugate to a vector field X' on a manifold M' if there is a diffeomorphism of class C^t of M onto M' which carries every oriented orbit of X into an oriented orbit of X'.

In other words: a diffeomorphism h of M onto M' is a conjugation from X to X' if the map $h \circ c$ is an integral curve of X' for every integral curve c of X (up to a parameter transformation preserving the sense of time). For a smooth h this is the case if there is a strictly positive function f on M' such that $h^T \circ X \circ h^{-1} = fX'$.

An analogous definition introduces the concept of conjugation for direction fields (up to orientation of orbits).

If the direction fields associated with two vector fields X and X' without singularities are C^t-conjugate, then the vector field X is either C^t-conjugate to the field X' or to the field -X' (the manifolds being assumed to be connected).

1.10. DEFINITION. A local diffeomorphism f of a manifold N is called C^t-conjugate to a local diffeomorphism f' of a manifold N' if there exists a diffeomorphism h of class C^t of N onto N' satisfying $f' \circ h = h \circ f$.

1.11. Remarks.

i) A conjugation carries singular points into singular points, and periodic orbits into periodic orbits, preserving all stability properties.

ii) The concepts of stability and of conjugation are independent of the choice of parameters for the orbits. We may therefore assume in the study of these questions, if necessary, that the occurring vector fields are complete.

1.12. <u>Exercises</u>.

i) If two diffeomorphisms are conjugate, so are their suspension vector fields.

ii) Let the notations of remark iii) of I-1.11 be used, and let X denote the vector field induced on $M \times S^1$ by the field $Y = Z(y,s) + \frac{\partial}{\partial s}$, with Z of class C^r and of period τ in s.

This field X is C^r-conjugate to the suspension vector field of the diffeomorphism h_τ of M.

In particular, the constant vector field with components α and 1 in $\mathbb{R}^2$ is invariant under translations. It determines a vector field on the torus $T^2 = \mathbb{R}^2/\mathbb{Z}^2$ which is analytically conjugate to the suspension field of the rotation through the angle α of the circle $S^1 = \mathbb{R}/\mathbb{Z}$.

2. <u>LINEAR DIFFERENTIAL EQUATIONS</u>

Let A be a map of class C^r (respectively analytic) of an open interval J of $\mathbb{R}$ into the space of $m \times m$ square matrices with real entries, and consider the linear differential equation on $\mathbb{R}^m$

$$\text{(E)} \qquad x' = A(t)x.$$

2.1. <u>Lemma</u>. The maximal integral curves of (E) are defined on all of J.

Proof. Let c be an integral curve of E defined on a bounded interval K whose closure is contained in J, and denote $k = \sup_{t \in K} \|A(t)\|$. Then we have $\|c'(t)\| \leq k\|c(t)\|$ for all $t \in K$. Hence c is bounded, and we apply lemma I-1.13 (cf. remark ii) of I-1.3). Q.E.D.

The set S of the solutions of (E) on J is a subspace of the vector space of all smooth maps of J into R^m. From the existence and uniqueness of the solutions we deduce the following result:

2.2. THEOREM. For k solutions $c_1, \ldots, c_k$ of (E) on J to be independent in S it is necessary and sufficient that there be a $t_o \in J$ such that the vectors $c_1(t_o), \ldots, c_k(t_o)$ are independent.

In this case the vectors $c_1(t), \ldots, c_k(t)$ are independent for all $t \in J$.

2.3. COROLLARY. The solution space S of (E) on J has dimension m.

2.4. A smooth map $t \longmapsto X(t)$ of J into the space of $m \times m$ square matrices is then called a fundamental matrix solution of (E) if the columns of X form m independent solutions of (E) on J. Such a map is a solution of the linear matrix differential equation $X' = A(t)X$. Hence we deduce:

i) if X and Y are two fundamental matrix solutions of (E) we have $X(t)X(t_o)^{-1} = Y(t)Y(t_o)^{-1}$ for all t and t_o in J;

ii) for a fundamental matrix solution X of (E) and a solution c of (E) on J we have $c(t) = X(t)X(t_o)^{-1}c(t_o)$ for all t and t_o in J.

For these reasons the matrix function $Z(t,t_o) = X(t)X(t_o)^{-1}$ which is independent of the choice of the fundamental matrix solution

X, is called the resolvent of (E).

This resolvent has the following properties (which resemble those indicated in remark iii) of I-1.11):

i) $Z(t_o,t_o) = I$;

ii) $Z(t_2,t_1)Z(t_1,t_o) = Z(t_2,t_o)$;

iii) $Z(t_1,t_o)^{-1} = Z(t_o,t_1)$;

iv) $\det Z(t_1,t_o) = \exp\int_{t_o}^{t_1} \mathrm{tr}A(s)ds$

(for $f(t) = \det Z(t,t_o)$ we find $f'(t)/f(t) = \mathrm{tr}A(t)$).

2.5. Example. For a constant matrix A the resolvent of (E) is $\exp(t-t_o)A$.

2.6. Remarks.

i) A similar theory of linear differential equations may be developed in $\mathbb{C}^m$, or even more generally, in a real or complex vector space of finite dimension.

ii) A linear transformation $y = S(t)x$ carries (E) into the linear differential equation $y' = B(t)y$, with

$$B(t) = S'(t)S(t)^{-1} + S(t)A(t)S(t)^{-1} .$$

For A and S constant matrices, B becomes also constant, and A and B are similar matrices.

2.7. Exercise. For a smooth map b of J into $\mathbb{R}^m$ the differential equation

$$\text{(F)} \qquad y' = A(t)y + b(t)$$

is called linear and non-homogeneous.

Under these conditions we have:

i) the set of all solutions of (F) on J is an affine space having S as its underlying vector space.

ii) For a fundamental matrix solution X(t) of (E) we look for a solution of (F) of the form $y(t) = X(t)c(t)$. The solution of (F) with initial condition $y(t_o) = y_o$ is thus found to be

$$y(t) = X(t)X(t_o)^{-1}y_o + X(t)\int_{t_o}^{t} X(s)^{-1}b(s)ds.$$

This procedure is known as "variation of the constants".

3. LINEAR DIFFERENTIAL EQUATIONS WITH CONSTANT COEFFICIENTS

3.1. Such equations are investigated by transforming A into its Jordan form [10].

Let F_- (respectively F_+ or F_o) denote the projections into $\mathbb{R}^m$ of the direct sum of the eigenspaces of A in $\mathbb{C}^m$ corresponding to those eigenvalues of strictly negative real parts (respectively strictly positive or zero). These subspaces, which are invariant under A, form a direct sum decomposition of $\mathbb{R}^m$. Choosing the coefficients $a_{i,i+1}$ of the Jordan form of A sufficiently small we can determine a scalar product on $\mathbb{R}^m$ with the following properties:

i) The subspaces F_-, F_+, and F_o are mutually orthogonal;

ii) $Ax.x < 0$ for every $x \in F_- - \{0\}$;

iii) $Ax.x > 0$ for every $x \in F_+ - \{0\}$;

iv) $Ax.x = 0$ for every $x \in F_o$ if the restriction of A to F_o is semi-simple.

The scalar square x.x is then a strict Liapounov function for E in O on the subspace F_-. Hence the origin is in F_- a singular and asymptotically ω-stable point in Liapounov's sense. Similarly the origin is in F_+ a singular and asymptotically α-stable point in Liapounov's sense. For this reason F_- and F_+ are respectively called the stable and unstable subspaces of the equation (E).

Finally, for a semi-simple restriction of A to F_o, the origin is in F_o a singular and ω- and α-stable point in Liapounov's sense. But now the stability is no longer asymptotic.

3.2. Equations in $\mathbb{R}^2$. We make a more detailed study of the case $m = 2$ and for an invertible matrix A. The equation (E) is then linearly conjugate to a linear differential equation $y' = By$, where the matrix B has one of the following forms:

$\begin{pmatrix} \lambda & 0 \\ 0 & \mu \end{pmatrix}$ or $\begin{pmatrix} \lambda & \frac{\lambda}{|\lambda|} \\ 0 & \lambda \end{pmatrix}$ for real eigenvalues λ and μ of A;

$\begin{pmatrix} \alpha & \beta \\ -\beta & \alpha \end{pmatrix}$ for eigenvalues $\alpha \pm i\beta$, $\beta \neq 0$ of A.

The following situations may then be distinguished:

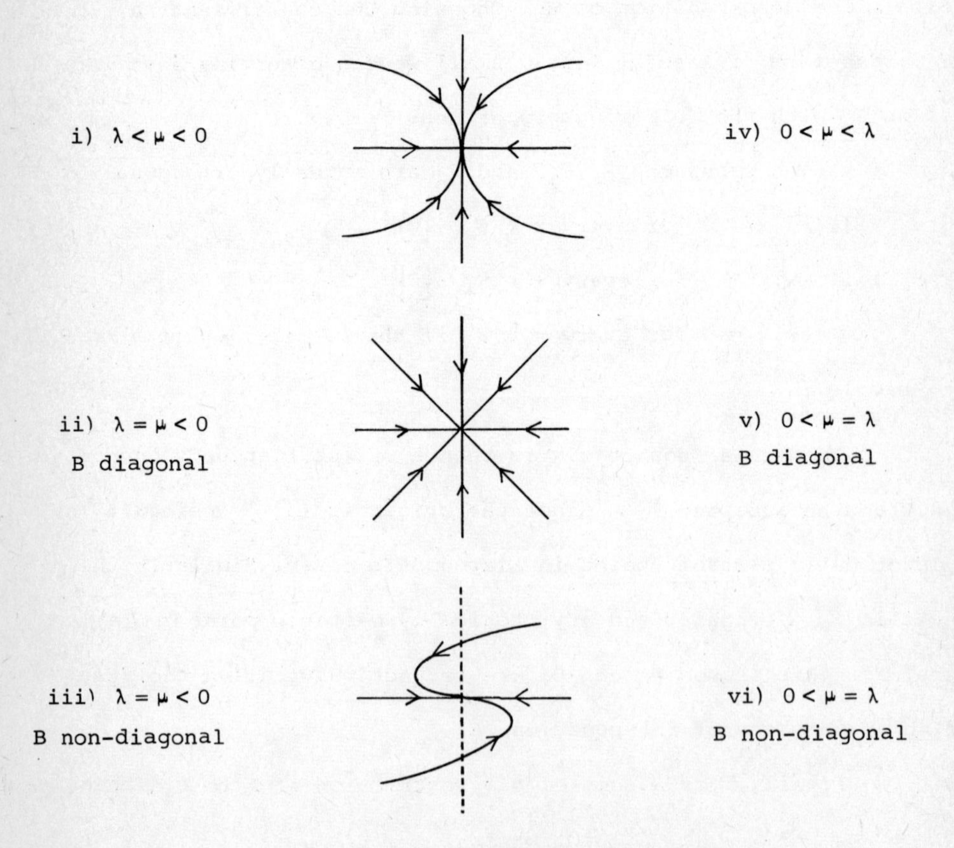

The latter cases iv), v), and vi) correspond to the former under a change of t into -t (i.e. under a change of orientation of the orbits).

vii) $\lambda < 0 < \mu$

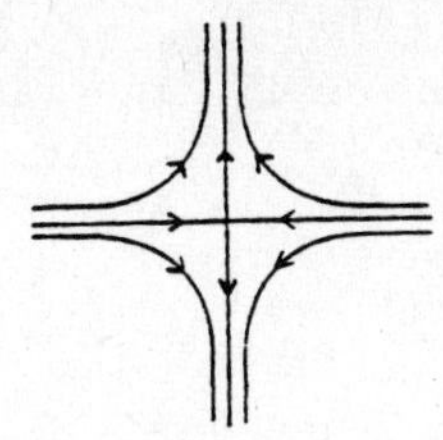

viii) $\alpha < 0$

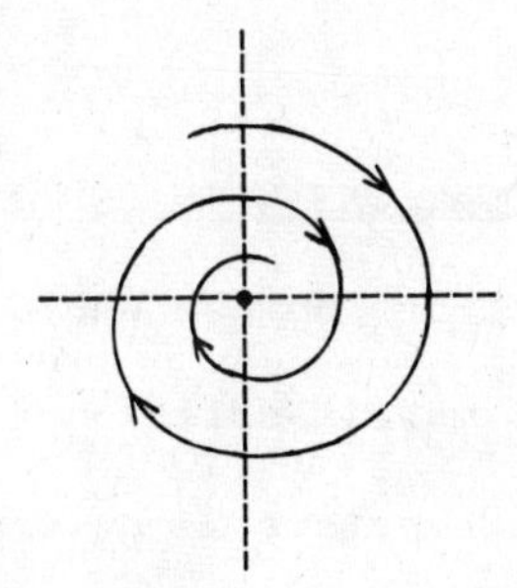

ix) $\alpha > 0$

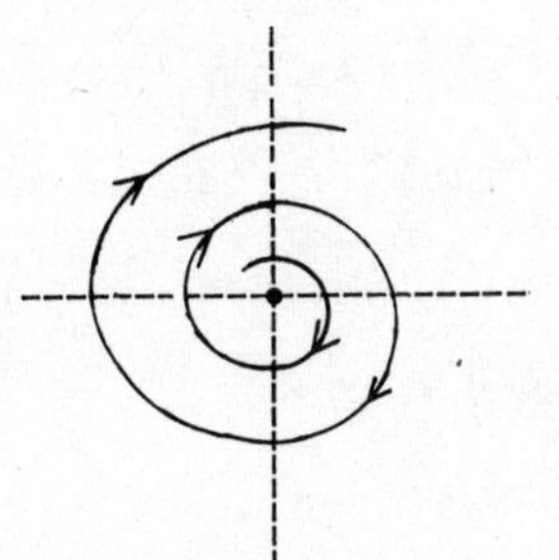

x) $\alpha = 0$

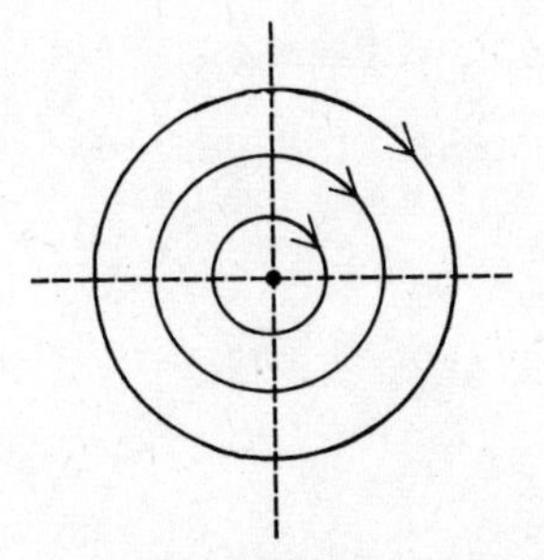

The cases viii), ix), and x) correspond to a positive β; those for $\beta < 0$ are obtained e.g. by interchanging x and -x.

The origin is now called

a <u>stable node</u> in the cases i), ii), and iii);

an <u>unstable node</u> for iv), v), and vi);

a <u>saddle point</u> for vii);

a <u>stable focus</u> in the case viii);

an <u>unstable focus</u> for case ix);

a <u>centre</u> for case x).

4. LINEAR DIFFERENTIAL EQUATIONS WITH PERIODIC COEFFICIENTS

4.1. Assume $A(t)$ to be defined on the whole of $\mathbb{R}$ and periodic with period τ. For a fundamental matrix solution X of (E) the function $X(t+\tau)$ is another one, hence there exists a constant matrix D satisfying $X(t+\tau) = X(t)D$ for all $t \in \mathbb{R}$.

Choose now a matrix B (possibly with complex entries) such that $e^{\tau B} = D$, and let $S(t) = e^{tB}X(t)^{-1}$. Then we have $S(t+\tau) = S(t)$, and the change of variables $y = S(t)x$ carries (E) into the linear differential equation $y' = By$.

Thus we have shown:

4.2. PROPOSITION (Theorem of Floquet). A linear differential equation (E) with periodic coefficients (of period τ) can be transformed into a linear differential equation with constant coefficients by a linear and periodic change of variables (of period τ).

4.3. If Y denotes another fundamental matrix solution of (E) then there is a constant matrix C such that $Y = XC$, and $Y(t+\tau) = Y(t)C^{-1}DC$. Hence the eigenvalues of the matrix D are invariants of the equation (E); they are called the <u>characteristic numbers</u> of (E).

Since $D = X(t)^{-1}X(t+\tau)$ the product of these characteristic numbers equals $\exp\int_0^\tau \mathrm{tr}A(s)ds$ (cf. 2.4).

The eigenvalues of the matrix B are, up to a factor $\frac{1}{\tau}$, logarithms of the characteristic numbers of (E); hence their real parts are independent of the choice of B. They are called the <u>characteristic exponents</u> of (E).

Let us consider now the zero solution of (E) as a periodic solution of the vector field $A(s)x + \frac{\partial}{\partial s}$ on the cylinder $\mathbb{R}^m \times (\mathbb{R}/\tau\mathbb{Z})$ (cf. remark iii) of I-1.11). Then we arrive at

4.4. <u>PROPOSITION</u>. If all characteristic numbers of (E) are in absolute value less (respectively greater) than 1, then the zero solution is asymptotically ω-stable (respectively α-stable) in Liapounov's sense.

4.5. <u>PROPOSITION</u>. If all characteristic numbers of (E) are simple and of absolute value 1, then the zero solution of (E) is ω- and α-stable in Liapounov's sense.

Indeed the change of variables $y = S(t)x$ of 4.1 is periodic, and therefore there is a constant $k > 0$ such that $\frac{1}{k} \leq \|S(t)\| \leq k$ for all $t \in \mathbb{R}$.

4.6. <u>Remarks</u>.

i) Letting $Z(t,t_0)$ denote the resolvent of (E) and choosing for X the fundamental matrix solution $Z(t,0)$ we obtain $D = Z(\tau,0)$. Hence (E) has a non-vanishing and periodic solution of period τ if and only if one of its characteristic numbers equals 1: for $v = c(0)$ this condition is indeed equivalent to $Dv = Z(\tau,0)v = v$.

ii) If (E) is considered as an equation of period 2τ then $X(t+2\tau) = X(t)D^2$. Hence in this case the matrix B may be chosen to be real.

iii) The similarity class of the matrix D is invariant under linear and periodic changes of variables (of period τ).

5. VARIATION FIELD OF A VECTOR FIELD

5.1. Let X be a vector field on the manifold M, and let (W,Φ) be a local flow of diffeomorphisms of M generated by X. Let $p: T(M) \rightarrow M$ denote the canonical projection, and V the open subset

$$\{(t,v) \in \mathbb{R} \times T(M) \mid (t,p(v)) \in W\} \qquad \text{of } \mathbb{R} \times T(M).$$

The "derivative of Φ with t kept constant" is a smooth map Ψ of V into T(M) whose restriction Ψ_t to $[\{t\} \times T(M)] \cap V$ is the tangent map to the restriction φ_t of Φ to $[\{t\} \times M] \cap W$.

Hence (V,Ψ) is a local flow of diffeomorphisms of T(M) whose germ along $\{0\} \times T(M)$ depends only on X. This local flow determines a vector field TX on T(M) with the following properties (cf. remark I-1.8):

i) the projection of every integral curve of TX is an integral curve of X.

ii) If $c: J \rightarrow M$ is an integral curve of X, then the curve $c': J \rightarrow T(M)$ is an integral curve of TX, because of $c'(t+s) = \varphi_t^T(c'(s))$.

iii) If $C(t,s) = c_s(t)$ denotes a smooth one-parameter family of integral curves of X, then the map $\gamma: t \mapsto \frac{\partial C}{\partial s}(t,s_o)$ is an integral curve of TX: for $C(t,s) = \varphi_{t-t_o}(C(t_o,s))$ and $\gamma(t) = \varphi^T_{t-t_o}(\gamma(t_o))$. Observe that the preceding case corresponds to the special family $C(t,s) = c(t+s)$.

For these reasons TX is called the <u>variation field</u> of the field X. (Note that the tangent map $X^T: T(M) \longrightarrow T(T(M))$ is not a vector field on T(M). If s denotes the canonical involution of T(T(M)) we have, however, $TX = s \circ X^T$ [15].)

5.2. <u>Remark</u>. The variation field TX of a complete vector field X is complete as well, and the flow generated by TX is then a one-parameter group of automorphisms of the tangent bundle T(M).

5.3. <u>Local Expression</u>. Let X be of class C^r, $r \geq 2$, and consider a chart of M. Locally we can write

$$\varphi_t(y) = y + tX(y) + o(t,y),$$

where $o(t,y)/t \rightarrow 0$ for $t \rightarrow 0$, uniformly in y on every compact subset.

Letting A(y) denote the linear map tangent to X in y we have

$$\varphi_t^T(y) = I + tA(y) + o'(t,y),$$

where $o'(t,y)/t \rightarrow 0$ as well for $t \rightarrow 0$.

For a tangent vector v at y we have therefore

$$TX(v) = (X(y), A(y)v).$$

More exactly: if the local expression of X is $\sum_i a_i \frac{\partial}{\partial y_i}$, and if $v_i = dy_i$, then the vector field TX corresponds to the differential equation

$$\left.\begin{aligned} y_i' &= a_i(y) \\ v_i' &= \sum_j \frac{\partial a_i}{\partial y_j}(y) v_j \end{aligned}\right\} \qquad i=1,2,\ldots,m.$$

This expression shows in particular that the field TX is of class C^{r-1} for X in class C^r.

5.4. <u>Singular Point</u>. If y is a singular point of X, then the tangent space $T_y(M)$ is invariant under TX, and this field induces on $T_y(M)$ a linear differential equation with constant coefficients. It is called the <u>variational equation</u> of X in y. The eigenvalues of this equation are called the <u>characteristic values</u> of X in y.

In the above chart this variational equation reads as follows:

$$v_i' = \sum_j \frac{\partial a_i}{\partial y_j}(y)v_j \ , \qquad i=1,2,\ldots,m.$$

5.5. <u>Periodic Orbits</u>. Let γ be a periodic orbit, of period τ, of X, and let y be a point of γ. The eigenvalues of the map $\varphi_\tau^T: T_y(M) \to T_y(M)$ are independent of the choice of the point y on γ; they are called the <u>characteristic numbers</u> of the periodic orbit γ.

One of these characteristic numbers equals 1 because of $\varphi_\tau^T X(y) = X(y)$; the m-1 others are called the <u>non-trivial</u> characteristic numbers.

If the map φ_τ^T preserves the orientation of $T_y(M)$ then the restriction E of T(M) to γ is a trivial bundle. A trivialisation h of E satisfying $h(0,s) = \varphi_s(y)$ then determines a vector field of the form $A(s)x + \frac{\partial}{\partial s}$ on the cylinder $\mathbb{R}^m \times (\mathbb{R}/\tau\mathbb{Z})$, or a linear differential equation (E) with periodic coefficients, of period τ, on $\mathbb{R}^m$. It is also called the <u>variational equation</u> of X along γ.

The map $t \mapsto h^{-1} \circ \varphi_t^T \circ h$ provides the fundamental matrix solution of (E) which coincides with the identity for $t = 0$. Hence, in this case the characteristic numbers of γ coincide with the characteristic numbers of (E) (cf. remark i) of 4.6).

5.6. <u>Examples</u>.

i) Let $X = (a_1,\ldots,a_m)$ denote a vector field on $\mathbb{R}^m$, and $c: \mathbb{R} \to \mathbb{R}^m$ a periodic integral curve of X, of period τ. Then the variational equation of X along the orbit γ corresponding to c is

$$v_i' = \sum_j \frac{\partial a_i}{\partial y_j}(c(t))v_j, \qquad i=1,2,\ldots,m.$$

Accordingly the product of the characteristic numbers of γ

(cf. 4.3 and 5.8) equals

$$\exp\int_0^T \Sigma_i \frac{\partial a_i}{\partial y_i}(c(t))\,dt = \exp\int_\gamma \operatorname{div} X\,dt \ .$$

ii) Let X be the suspension field of a diffeomorphism f of a manifold N (cf. I-4.10), and y be a fixpoint of f. The orbit γ of X passing through y is then periodic with period 1.

The canonical decomposition of the tangent bundle of the product $N \times \mathbb{R}$ is invariant under the action of $\frac{\partial}{\partial s}$. It determines then a decomposition of T(M) into the Whitney sum of the subbundle of rank 1 generated by X and of a subbundle of rank m-1 which is invariant under X. (Its restriction to the submanifold N of M is the tangent bundle T(N).

The restriction of the map $\varphi_1^T: T_y(M) \longrightarrow T_y(M)$ to the subspace $T_y(N)$ coincides with the tangent map to f at y. Hence the non-trivial characteristic numbers of γ are the eigenvalues of the map $f_y^T: T_y(N) \longrightarrow T_y(N)$.

5.7. <u>Exercise</u>. Let X denote the vector field on the cylinder $\mathbb{R}^2 - \{0\}$ corresponding to a differential equation $\frac{d\rho}{d\theta} = f(\rho,\theta)$.

If $f(1,\theta)$ identically vanishes then the unit circle is a periodic orbit of X of period 2π, whose non-trivial characteristic number equals $\exp\int_0^{2\pi} \frac{\partial f}{\partial \rho}(1,\theta)\,d\theta$.

5.8. <u>PROPOSITION</u>. The product of the characteristic numbers of a periodic orbit γ of X equals $\exp\int_\gamma \operatorname{div} X\,dt$.

(Let M be an orientable manifold and ω a volume form on M. Then the <u>divergence</u> of the vector field X with respect to ω is the smooth function on M determined by the relation $L_X\omega = (\operatorname{div} X)\,\omega$.)

<u>Proof</u>. The relation $\varphi_t^* \omega = f_t \omega$ determines a smooth function $f_t(x)$ on $M \times R$ for which we find

$$\frac{\partial f_t}{\partial t}\, \omega = \frac{d}{dt}(\varphi_t^* \omega) = \varphi_t^*(L_X \omega) = (\operatorname{div} X \circ \varphi_t) f_t \omega .$$

Thus

$$f_t(x) = \exp \int_0^t \operatorname{div} X(\varphi_t(x))\, dt .$$

If, however, τ is the period of γ and y is one of its points, the product of the characteristic numbers of γ is equal to $f_\tau(y)$. Q.E.D.

5.9. <u>The case of surfaces</u>. Let γ denote a periodic orbit of a vector field X without singularity on an orientable surface M. (This is the general situation in the neighbourhood of a periodic two-sided orbit.)

For a volume form ω on M the Pfaffian form $\alpha = i_X \omega$ has no singularities, and it is possible to find another Pfaffian form $\bar{\alpha}$ satisfying $\omega = \alpha \wedge \bar{\alpha}$ and $\bar{\alpha}(X) = -1$. (For a Pfaffian non-singular form α a differential form β satisfying $\alpha \wedge \beta = 0$ is divisible by α.) We thus have $d\alpha = \alpha \wedge (\operatorname{div} X\, \bar{\alpha})$, and the non-trivial characteristic number of γ is $\exp(-\int_\gamma \operatorname{div} X\, \bar{\alpha})$.

Assume now that β is a second Pfaffian form without singularities on M satisfying $\beta(X) = 0$, and $\bar{\beta}$ a Pfaffian form such that $d\beta = \beta \wedge \bar{\beta}$. Then we have the following relations:

$\beta = f\alpha$ where f is a nowhere vanishing function on M;

$$d\beta = \beta \wedge (\operatorname{div} X\, \bar{\alpha} - \frac{df}{f}),$$

$$\bar{\beta} = \operatorname{div} X\, \bar{\alpha} - \frac{df}{f} + g\beta, \text{ where } g \text{ is defined on } M,$$

$$\int_\gamma \bar{\beta} = \int_\gamma \operatorname{div} X\, \bar{\alpha} .$$

This leads to the following result:

5.10. <u>PROPOSITION</u>. Let E be an oriented direction field on a surface M defined by a Pfaffian form α without singularity, and let $\bar{\alpha}$ be a

Pfaffian form on M satisfying $d\alpha = \alpha \wedge \bar{\alpha}$. Then the characteristic number of a periodic orbit γ of E equals $\exp(-\int_\gamma \bar{\alpha})$.

5.11. Exercise. Let E denote the direction field on the cylinder $C = \mathbb{R}^2 - \{0\}$ which in polar coordinates is determined by the Pfaffian form $\alpha = d\rho - f(\rho,\theta)d\theta$, where f is periodic in θ with period 2π. E will be oriented in the sense of increasing θ.

i) If $f(\rho,\theta)$ is an affine function of ρ then the field E either has only periodic orbits, or possesses at most one periodic orbit.

ii) If f''_ρ is strictly positive on C then the field E has at most three periodic orbits, viz. at most one for which the characteristic exponent is strictly negative, respectively zero, or strictly positive (cf. III-1 and III-2).

6. BEHAVIOUR NEAR A SINGULAR POINT

6.1. Let y be a singular point of the vector field X. Considering a chart of M around y we can reduce the investigation to the case where X is a vector field on the space $\mathbb{R}^m$ which vanishes at the origin. Letting A denote the Jacobian matrix of X at 0 we can write $X(x) = Ax+o(x)$. The linear part of X at 0 corresponds thus to its variational equation in this point.

Let now f be a diffeomorphism of class C^1 of $\mathbb{R}^m$ having the origin as a fixpoint, and let S be the Jacobi matrix of f at 0. The vector field $Y = f^T \circ X \circ f^{-1}$ (which in general is only of class C^0) can be expressed as $Y(x) = SAS^{-1}x + o(x)$. Hence we have:

6.2. PROPOSITION. The characteristic values of X in a singular point are invariant under conjugations of class C^1.

6.3. Remark. If none of these characteristic values vanishes then y is an isolated singular point of X, and y is said to be a non-degenerate singular point of X.

6.4. THEOREM. If the characteristic values of X at the singular point y have strictly negative (respectively strictly positive) real parts, then y is an asymptotically ω-stable (respectively α-stable) singular point in Liapounov's sense. Furthermore the field X is topologically conjugate near y to the linear vector field $Y(x) = -x$ (respectively $Z(x) = x$) on $\mathbb{R}^m$.

Proof. As above we may write $X(x) = Ax + o(x)$ in $\mathbb{R}^m$. If the eigenvalues of A have all strictly negative real parts then there exists a scalar product on $\mathbb{R}^m$ and a number $\varepsilon > 0$ such that the corresponding scalar square becomes a strict Liapounov function for X at 0 on the open ball B of centre 0 and radius ε (cf. 3.1). Hence the origin will be an asymptotically ω-stable singular point in Liapounov's sense (cf. proposition 1.6).

Denoting by (φ_t) the flow generated by X we obtain a homeomorphism h of the open unit ball D of center 0 onto the ball B by setting

$$h(tx) = \varphi_{-\mathrm{Log}\, t}(\varepsilon x) \text{ for } x \in S^{m-1} \text{ and } t \in (0,1),$$
$$h(0) = 0,$$

and this homeomorphism is a conjugation from $Y|D$ to $X|B$.

The case of strictly positive real parts of the eigenvalues of A is treated by changing t into -t. Q.E.D.

6.5. Remarks.

i) In general the above conjugation is not of class C^1 (proposition 6.2). It is, however, a diffeomorphism except at the origin.

ii) Theorem 6.4 shows that the field X is topologically conjugate to its variational equation near y, provided all characteristic values of X in y have strictly negative (respectively positive) real parts. This is actually even true for a hyperbolic singular point, i.e. if none of the characteristic values of X at y is purely imaginary [11].

6.6. Exercises.

i) Let X and Y be two linear vector fields on $\mathbb{R}^m$ with invertible matrices and having no purely imaginary eigenvalues. Then the fields X and Y are topologically conjugate if and only if their stable subspaces (and hence also their unstable subspaces) have the same dimension. In particular there exist four topological types of singularities for linear fields in $\mathbb{R}^2$: the stable and unstable nodes (or foci), the saddles, and the centers (cf. theorem III-3.12).

ii) The vector field $(x,y) \longmapsto (y-x^3, -x-y^3)$ in $\mathbb{R}^2$ is not topologically conjugate to its linear part near the origin. On the other hand it is conjugate to the linear field $(x,y) \longmapsto (-x,-y)$.

iii) In the plane the vector field with the following components is of class C^1:

$$x(x^2+y^2)^3 \sin\frac{1}{x^2+y^2} - y, \quad y(x^2+y^2)^3 \sin\frac{1}{x^2+y^2} + x.$$

Near the origin it is, however, not conjugate to any analytic vector field (cf. III-3).

7. BEHAVIOUR NEAR A PERIODIC ORBIT

7.1. Poincaré's map. Let γ denote a periodic orbit of period τ of a vector field X of class C^r, and let y be a point of γ, and N a submanifold of codimension 1 of M which is transverse to X and satisfies $N \cap \gamma = \{y\}$. The maximal flow generated by X is denoted by $\Phi = (\varphi_t)$.

Introduce a distinguished chart $(y_1,\ldots,y_m)$ on an open neighbourhood Ω of y in which N is defined by $y_1 = 0$. Then the equation $\Phi_1(t,0,y_2,\ldots,y_m) = 0$ may be solved for t near τ, yielding a numerical function α on an open neighbourhood U of y in N which has the same class of differentiability as X and satisfies the following properties:

i) $\alpha(y) = \tau$;

ii) $\Phi(\alpha(z),z) \in N$ for all $z \in U$.

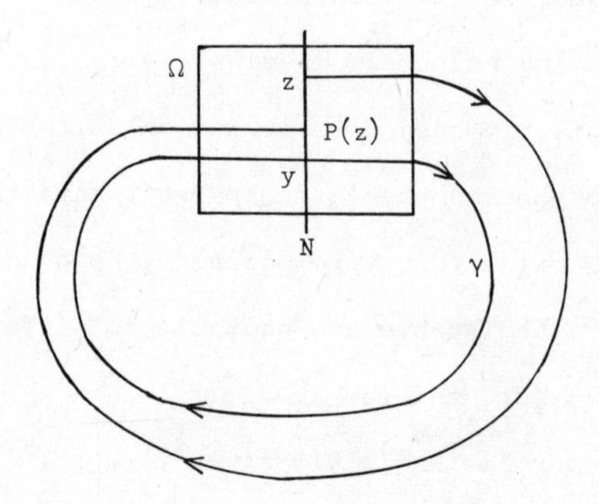

This function $\alpha(z)$ may therefore be interpreted as the time of first return to N of the integral curve of X passing through z.

The map P: $z \mapsto \Phi(\alpha(z),z)$ leads then from U to N, leaves y fixed, and has the same smoothness as X. It is called the <u>map of Poincaré</u> for γ (and the transverse submanifold N).

Poincaré's map is invertible near y with the inverse map $z \mapsto \Phi(-\alpha(z),z)$. It preserves the orientation of N if and only if the restriction E of the tangent bundle T(M) to γ is trivial (which is always true for an orientable M). Its periodic points correspond to periodic orbits of X.

7.2. Example. Let X be the suspension field of a diffeomorphism of a manifold N, and let γ be the periodic orbit of X corresponding to a fix-point y of f. Then the map f near y is a Poincaré map for γ.

We are going to show now how a Poincaré map for γ characterizes the behaviour of the orbits of X near γ.

For this purpose we consider a second vector field X' of class C^r on a manifold M', and a periodic orbit γ' of period τ' of X'. We define analoguous concepts for γ' as those introduced for γ in 7.1 denoting corresponding elements of γ' by the same but primed letter.

7.3. THEOREM. Let $t \leqslant r$. For the fields X and X' to be C^t-conjugate in the neighbourhoods of γ and γ' it is necessary and sufficient that for their Poincaré maps the same condition holds near y and y'.

Proof. Assume first P and P' conjugate, and choose U in $N \cap \Omega$ and U' in $N' \cap \Omega'$ in such a way that there exists a diffeomorphism f of N into N' satisfying $f(y) = y'$ and $P' \circ f = f \circ P$ on U.

Choose also Ω, Ω', U, and U', as well as a number $\varepsilon > 0$ less than $\tau/10$ and $\tau'/10$ such that we have

i) the open set Ω (respectively Ω') is defined by $|y_i| < 4\varepsilon$ (repectively $|y_i'| < 4\varepsilon$);

ii) Φ (respectively Φ') is defined on $(-2\varepsilon, \tau+\varepsilon) \times \Omega$ (respectively on $(-2\varepsilon, \tau'+3\varepsilon) \times \Omega'$, and is injective on $[-\varepsilon, \tau-2\varepsilon] \times U$ (respectively on $[-\varepsilon, \tau'-2\varepsilon] \times U'$);

iii) $\tau-\varepsilon < \alpha(z) < \tau+\varepsilon$ for every $z \in U$ (respectively $\tau'-\varepsilon < \alpha'(z') < \tau'+\varepsilon$ for every $z' \in U'$).

Then we construct a map β of class C^t of $\mathbb{R}\times U$ into $\mathbb{R}$ with the following properties:

i) the function $t\mapsto\beta(t,z)$ is strictly increasing for every $z\in U$;

ii) $\beta(t,z) = t$ for $t\leq 5\varepsilon$;

iii) $\beta(t,z) = t+\alpha'(f(z))-\alpha(z)$ for $t\geq\alpha(z)-3\varepsilon$.

The map $(t,z)\mapsto\Phi'(\beta(t,z),f(z))$ then induces a diffeomorphism of class C^t of the open neighbourhood $W=\Phi((-\varepsilon,\tau+\varepsilon)\times U)$ of γ onto an open neighbourhood W' of γ' which carries γ into γ' and relates $X|W$ to $X'|W'$ by conjugation.

Conversely assume that h is a diffeomorphism of an open neighbourhood W of γ onto an open neighbourhood W' of γ' transforming γ into γ' and inducing a conjugation of $X|W$ into $X'|W'$.

Choosing W and W' sufficiently small and composing, if necessary, with a diffeomorphism φ_t we may suppose that $h(y) = y'$ and $h(\Omega)\subset\Omega'$.

Let then p' denote the projection of Ω' on N', parallel to the y_1'-axis. Then the composition $f=p'\circ h$ is a diffeomorphism of an open neighbourhood $V\subset U$ of y in N onto an open neighbourhood $V'\subset U'$ of y' in N' carrying y into y' and conjugating P into P' near these points. Q.E.D.

7.4. <u>Remarks</u>. Quite often the conjugation f of P into P' is of class C^0 and is smooth except at y and y' (cf. e.g. proposition 7.7 and 7.10). The conjugation of X to X' constructed above is then a diffeomorphism outside of γ and γ'.

7.5. <u>COROLLARY</u>. The class of C^r-conjugations of the germs of Poincaré maps for γ depends only on γ.

In particular we recognize as in 6.1 that the eigenvalues of the tangent map $P^T\colon T_y(N) \to T_y(N)$ of a Poincaré map for γ depend only on γ. These eigenvalues are the non-trivial characteristic numbers of γ (cf. 5.5) for we have indeed near y

$$\Phi(\tau,y_1,\dots y_m) = (\tau+y_1-\alpha(y_2,\dots,y_m),P_2(y_2,\dots,y_m),\dots,P_m(y_2,\dots y_m)).$$

7.6. <u>THEOREM</u>. Let the non-trivial characteristic numbers of γ all be strictly less (respectively strictly greater) than 1 in absolute value. Then the vector field X is topologically conjugate near γ to the suspension field of the tangent map P_y^T of the Poincaré map P for γ at y.

In view of the above results it is sufficient to check the following proposition:

7.7. <u>PROPOSITION</u>. Let f be a local diffeomorphism of $\mathbb{R}^m$ defined in a neighbourhood of 0 and satisfying $f(0) = 0$, and let the eigenvalues of the Jacobi matrix $A = Df(0)$ of f at 0 be strictly less than 1 in absolute value. Then the diffeomorphism f is topologically conjugate near 0 to its linear tangent map $x \mapsto Ax$ at 0.

<u>Proof</u>. Writing $f(x) = Ax + o(x)$ we may choose a scalar product on $\mathbb{R}^m$, and three positive numbers α,β and ϵ satisfying the following properties:

i) $0 < 4(1-\beta) < \alpha < \beta < 1$;

ii) $\alpha\|x\| \leqslant \|Ax\| \leqslant \beta\|x\|$ for every $x \neq 0$;

iii) $\|o(x)\| < (1-\beta)\|x\|$ and $\|Do(x)\| < 1-\beta$ for $\|x\| \leqslant \varepsilon$.

Let then θ denote a smooth function on $\mathbb{R}^m$ contained between 0 and 1, equalling 1 for $\|x\| \leqslant \frac{\varepsilon}{2}$, 0 for $\|x\| \geqslant \varepsilon$, and satisfying

$\|D\theta(x)\| \leqslant \frac{\varepsilon}{3}$ for every x. Then the map $g: x \longmapsto Ax + \theta(x)o(x)$ is a diffeomorphism of $\mathbb{R}^m$ coinciding with f for $\|x\| \leqslant \frac{\varepsilon}{2}$ and with the linear map $x \longmapsto Ax$ for $\|x\| \geqslant \varepsilon$. Moreover, g is a contraction.

We now obtain a bijection of $\mathbb{R}^m$ onto itself by letting $h(0)=0$, and $h(x) = g^r(A^{-r}x)$ for $x \neq 0$, where r is a positive integer chosen such that we have $\|A^{-r}x\| > \varepsilon$. This map h has the following properties:

i) It is a diffeomorphism of $\mathbb{R}^m - \{0\}$ onto itself.

ii) It is continuous at the origin: for a sequence x_p tending towards 0 and for the numbers $r_p = \inf(r \geqslant 0 \mid \|A^{-r}x_p\| > \varepsilon)$ we obtain $\|A^{-r_p}x_p\| \leqslant \frac{\varepsilon}{\alpha}$, $\lim_{p\to+\infty} r_p = +\infty$, $\lim_{p\to+\infty} g^{r_p}(A^{-r_p}x_p) = 0$, since g is a contraction.

iii) h^{-1} is a conjugation to the linear map $x \longmapsto Ax$ near 0 because of $g \circ h = h \circ A$. Q.E.D

7.8. <u>COROLLARY</u>. If the non-trivial characteristic numbers of the periodic orbit γ are all strictly less than 1 in absolute value (respectively greater than 1), then the orbit γ is asymptotically ω-stable (respectively α-stable) in Liapounov's sense.

7.9. <u>Remark</u>. Theorem 7.6 remains true even for a <u>hyperbolic</u> periodic orbit, i.e. a periodic orbit having no non-trivial characteristic number of absolute value 1 [11].

7.10. <u>The case of surfaces</u>. Let γ be a periodic orbit of a vector field X on a surface M. Then a Poincaré map P for γ may be interpreted as a real and monotonic function on a neighbourhood of 0 in $\mathbb{R}$ vanishing at the origin. The non-trivial characteristic number of γ equals the derivative $P'(0)$.

Assume first P to be increasing, which is true in particular for an orientable surface. On a sufficiently small neighbourhood and on both sides of 0 P may exhibit one of the four following types of behaviour:

i) P is a contraction

(If P is an increasing contraction on the interval $[0,\alpha]$ then it is topologically conjugate to the homothety $x \mapsto \frac{1}{2}x$ (cf. proposition A.1));

ii) P is an expansion (i.e. P^{-1} is a contraction);

iii) P is the identity map;

iv) P differs from the identity, but the origin is the limit of non-vanishing fixpoints of P.

These four possibilities lead to the following types of behaviour of the orbits of X close to γ and lying on the "considered side":

i) γ is asymptotically ω-stable in Liapounov's sense;

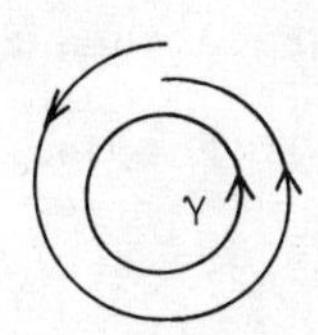

ii) γ is asymptotically α-stable in Liapounov's sense;

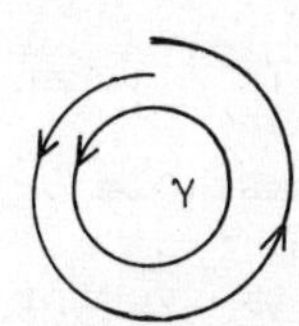

iii) γ is (ω- and α-) stable in Liapounov's sense, and all orbits close to γ are periodic;

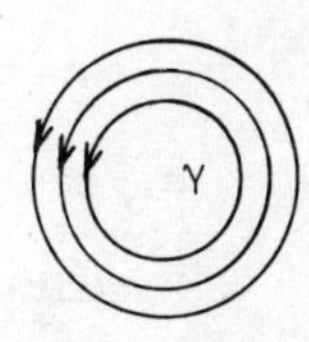

iv) γ is (ω- and α-)stable in Liapounov's sense, and it is a limit of periodic orbits, which are isolated if the non-vanishing fixpoints of P are isolated, and alternately asymptotically ω-stable and α-stable if $P(x)-x$ changes its sign at each of these fixpoints.

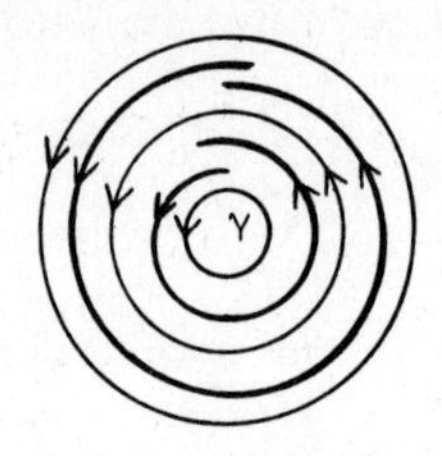

These four types can form combinations on the two sides of γ giving sixteen types of behaviour. Among them we may distinguish the following two cases:

the <u>stable limit cycle</u> if P is a contraction on both sides of 0;

the <u>unstable limit cycle</u> if P is an expansion on both sides of 0.

We have then the following properties:

i) If the derivative $P'(0)$ is strictly less than 1 (respectively strictly greater than 1), then γ is a stable (respectively unstable) limit cycle;

ii) if X is an analytic vector field, then γ is a limit cycle (stable or unstable or "mixed"), or all orbits close to γ are periodic;

iii) two vector fields are topologically conjugate in the neighbourhoods of stable, unstable, or mixed limit cycles.

Consider now the case of a decreasing P. This implies that M is non-orientable, and that γ has an open neighbourhood U which is diffeomorphic to a Moebius strip. We reduce this case to a periodic orbit $\tilde{\gamma}$ on the two-sheeted orientation covering space of U having P^2 as Poincaré map, and such that the orbits exhibit the same behaviour on either side of $\tilde{\gamma}$. There are thus four possibilities:

i) γ is a stable limit cycle (which occurs e.g. for $|P'(0)| < 1$);

ii) γ is an unstable limit cycle (which arises e.g. if $|P'(0)| > 1$);

iii) all orbits close to γ are periodic;

iv) γ is the limit of periodic orbits with increasing Poincaré maps.

In the last two cases the period of the orbits close to γ but distinct from γ approaches twice the period of γ.

From these considerations follows finally the following interesting result:

7.11. PROPOSITION. For a surface a limit set which contains a periodic orbit reduces to this periodic orbit.

7.12. Exercise. Produce a counterexample to this last statement in higher dimensions.

APPENDIX: CONJUGATION OF CONTRACTIONS IN $\mathbb{R}$

Local diffeomorphisms of $\mathbb{R}$ which are contractions near the origin will be studied in this appendix, because of their role in the theory of foliations.

A.1. PROPOSITION. An increasing homeomorphism f which is a contraction on the interval $[0,1]$ is topologically conjugate on this interval to the homothety $t \mapsto \frac{1}{2}t$.

Proof. Let h denote an increasing homeomorphism of the interval $[f(1),1]$ onto the interval $[\frac{1}{2},1]$. h may be extended to a homeomorphism of $[0,1]$ onto itself by defining $h(0) = 0$ and $h(t) = \frac{1}{2^n} h(f^{-n}(t))$ for

$t \in [f^{n+1}(1), f^n(1)]$. We thus obtain a topological conjugation of f to the homothety $t \mapsto \frac{1}{2}t$. Q.E.D.

A.2. Remarks.

i) If the above homeomorphism f is a diffeomorphism of class C^1 except for the origin, and if h is chosen as an increasing diffeomorphism of $[f(1),1]$ onto $[\frac{1}{2},1]$ satisfying $h'(1) = 2f'(1)h'(f(1))$, then the preceding construction leads to a conjugation which is a C^1 diffeomorphism save at the origin (cf. exercise A.5).

More generally for f of class C^r, $1 \leq r \leq +\infty$, except for the origin, one can construct such a conjugation with the same smoothness except at the origin.

ii) As in A.1 it can be shown that two increasing homeomorphisms f and g of the interval $[0,1]$ onto itself satisfying $f(t) < t$ and $g(t) < t$ for t different from 0 and 1 are topologically conjugate by an increasing homeomorphism of $[0,1]$ onto itself.

If in addition f and g are C^1 diffeomorphisms of the open interval $(0,1)$ it is possible to find a conjugation from f to g with the same property.

iii) Let f and g be two local diffeomorphisms of class C^1 of $\mathbb{R}$ with the origin as a fixpoint. If f and g are conjugate near 0 by a C^1 diffeomorphism keeping the origin fixed then $f'(0) = g'(0)$.

According to [9] we have:

A.3. THEOREM. Let f be an increasing diffeomorphism of class C^2 leaving the origin fixed and having on the interval $[0,1]$ a derivative strictly between 0 and 1. Then there exists on this interval a unique conjugation of class C^1 of f to the homothety $t \mapsto f'(0)t$.

<u>Proof</u>. Let the derivative of f at 0 be called a, and let h be a C^1 conjugation on $[0,1]$ from f to the homothety $t \mapsto at$. Then we have $a^n h(t) = h(f^n(t))$ for every integer $n \geq 0$. Hence the sequence

$$\frac{a^n}{f^n(t)} h(t) = \frac{h(f^n(t))}{f^n(t)},$$

$n \geq 0$ and $t \neq 0$, has a limit independent of t and equalling the derivative $b = h'(0)$. For every t in $[0,1]$ we thus have $h(t) = b \lim_{n \to +\infty} \frac{f^n(t)}{a^n}$. This proves the uniqueness of the conjugation (note that the assumption of smoothness of f was not used so far).

We will show now that the sequence $\frac{f^n}{a^n}$, $n \geq 0$, converges uniformly on the interval $[0,1]$ towards a C^1 conjugation from f to the homothety $t \mapsto at$.

Since $f(t) = at(1+O(t))$ we may write

$$\frac{f^n(t)}{a^n} = t \prod_{k=1}^{n} \frac{f^k(t)}{af^{k-1}(t)} = t \prod_{k=1}^{n} [1+O(f^{k-1}(t))].$$

Letting $A = \sup_{[0,1]} f'(t)$, we have, however, $f^k(t) \leq A^k t$ for every $t \in [0,1]$ and every $k \geq 1$. Hence the infinite product $\prod_{k=1}^{\infty} [1+O(f^{k-1}(t))]$ converges uniformly on $[0,1]$, and the same holds for the sequence $\frac{f^n}{a^n}$. The limit of this sequence then becomes a continuous function k satisfying $k(0) = 0$.

In an analoguous way the derived sequence $\frac{(f^n)'}{a^n} = \frac{1}{a^n} \prod_{k=0}^{n-1} f' \circ f^k$ is shown to converge uniformly on $[0,1]$, as well as the inverse sequence $\frac{a^n}{(f^n)'}$. Hence the map $h: t \mapsto \frac{k(t)}{k(1)}$ is a C^1-diffeomorphism of the interval $[0,1]$ onto itself. Finally we have

$$h[f(t)] = \frac{1}{k(1)} \lim_{n \to +\infty} \frac{f^{n+1}(t)}{a^n} = ah(t). \qquad \text{Q.E.D.}$$

A.4. <u>COROLLARY</u>. Under the assumptions of theorem A.3, let g denote a C^1-diffeomorphism which leaves the origin fixed and commutes with f on the interval $[0,1]$. Then the conjugate map $h \circ g \circ h^{-1}$ of g by h is a homothety.

A C^1-map φ commuting with a homothety $t \mapsto at$, $0 < a < 1$, is indeed itself a homothety, because of $\varphi(a^n t) = a^n \varphi(t)$ we have $\varphi(0) = 0$ and $\varphi'(t) = \varphi'(a^n t) = \varphi'(0)$.

A.5. Exercise. The map $f: t \mapsto at(1 + \frac{1}{\mathrm{Log}|t|})$, $0 < a < 1$, is a local diffeomorphism of class C^1 of $\mathbb{R}$ which is a contraction near the origin. But there does not exist any C^1-conjugation near 0 from f to the homothety $t \mapsto at$.

A.6. Remarks.

i) Under the assumptions of theorem A.3 one can show that for f of class C^r, $2 \leq r \leq +\infty$, the conjugation h is of class C^{r-1} [9].

ii) Theorem A.3 obviously remains no longer true if f has a derivative equal to 1 at 0. The following theorem, which is essentially due to N. Koppell [7], is, however, a generalisation of corollary A.4 to this situation.

A.7. THEOREM. Let f denote an increasing diffeomorphism of class C^2 which is a contraction on the interval $[0,1]$ and whose derivative at 0 equals 1. Then there exists an increasing homeomorphism h of the interval $[0,1]$ onto itself having the following properties:

i) h is a C^1diffeomorphism except for the origin;

ii) if g is an increasing C^1-diffeomorphism leaving the origin fixed and commuting with f on the interval $[0,1]$ then its conjugate $h \circ g \circ h^{-1}$ by h is a homothety.

In particular this homeomorphism h carries f into a homothety by conjugation.

The proof of this theorem makes use of the following lemma:

A.8. LEMMA. Under the assumptions of theorem A.7 the sequence $\frac{(f^n)'}{(f^n)'(1)}$ as well as the inverse sequence $\frac{(f^n)'(1)}{(f^n)'}$ converge uniformly on every compact interval $[a,1]$, $0 < a \leqslant 1$.

Proof. Let A (respectively B) denote the least upper bound of $1/f'(t)$ (respectively of $|f''(t)|$) on the interval $[0,1]$, and j a positive integer satisfying $f^j(1) \leqslant a$. In view of the relation

$$\frac{(f^n)'(t)}{(f^n)'(1)} = \prod_{k=0}^{n-1} \frac{f'(f^k(t))}{f'(f^k(1))} ,$$

it will be sufficient to show the uniform convergence on $[a,1]$ of the series

$$\sum_0^{\infty} \left| \frac{f'(f^k(t))}{f'(f^k(1))} - 1 \right| .$$

But we have

$$\begin{aligned} \left| \frac{f'(f^k(t))}{f'(f^k(1))} - 1 \right| &\leqslant AB|f^k(t) - f^k(1)| \\ &\leqslant AB|f^k(a) - f^k(1)| \\ &\leqslant AB|f^{k+j}(1) - f^k(1)| \\ &\leqslant AB\sum_{i=1}^{j} |f^{k+i}(1) - f^{k+i-1}(1)| . \end{aligned}$$

We conclude

$$\begin{aligned} \sum_{k=p}^{q} \left| \frac{f'(f^k(t))}{f'(f^k(1))} - 1 \right| &\leqslant AB \sum_{k=p}^{q} \sum_{i=1}^{j} |f^{k+i}(1) - f^{k+i-1}(1)| \\ &\leqslant AB \sum_{i=1}^{j} |f^{q+i}(1) - f^{p+i-1}(1)| \\ &\leqslant jABf^p(1). \end{aligned}$$

Q.E.D.

The limit $k(t)$ of this sequence is a continuous and strictly positive function on the interval $(0,1]$.

Proof of theorem A.7. Let g denote an increasing C^1-diffeomorphism leaving the origin fixed and commuting with f on the interval $[0,1]$. Differentiating the identities $f^n \circ g = g \circ f^n$, $n \geqslant 0$, we obtain the relations $$(f^n)'(g(t))g'(t) = g'[f^n(t)](f^n)'(t).$$

Passing to the limit this yields $k(g(t))g'(t) = g'(0)k(t)$.

Letting K denote the indefinite integral of k(t) which vanishes for t=1, we have $K[g(t)] = g'(0)K(t) + K[g(1)]$; in particular $K[f(t)] = K(t) + a$, with $a = K[f(1)] < 0$.

The map K is therefore a C^1-diffeomorphism of the interval $(0,1]$ onto the interval $(-\infty,0]$ which relates f to the translation $t \mapsto t+a$ by conjugation, and g to the affine map $t \mapsto g'(0)t + K[g(1)]$. Since f and g commute, however, this implies that g'(0) equals 1.

Under these circumstances the map $h: t \mapsto e^{K(t)}$ may be extended to an increasing homeomorphism of the interval $[0,1]$ onto itself and satisfying the listed conditions. Q.E.D.

A.9. <u>COROLLARY</u> (lemma of N. Koppell [7]). Let f denote an increasing diffeomorphism of class C^2 which is a contraction on the interval $[0,1]$, and let g be an increasing diffeomorphism of class C^1 leaving the origin fixed and commuting with f on this interval. If g has a fixpoint in $(0,1]$ then it coincides with the identity map on $[0,1]$.

Since g has then fixpoints arbitrarily close to 0 this corollary is an immediate consequence of corollary A.4 and of theorem A.7.

A.10. <u>COROLLARY</u>. Let f be an increasing C^2-diffeomorphism of the interval $[0,1]$ onto itself having no fixpoint on the open set $(0,1)$, and let g be an increasing C^1-diffeomorphism of the interval $[0,1]$ commuting with f. If g has a fixpoint in $(0,1)$ then it is the identity map on $[0,1]$.

Proof. We may assume f to be a contraction on $[0,1)$. If t is then a fixpoint of g in $(0,1)$, the points $t_n = f^n(t)$, $n \in Z$ (which accumulate at 0 and 1) are fixpoints of g as well, and g is determined by its restriction to one of the intervals $[t_{n+1}, t_n]$. The conclusion follows from corollary A.9. Q.E.D.

A.11. Remarks.

i) One can show that the homeomorphism h constructed in the proof of theorem A.7 is of class C^{r-1} on $(0,1]$ if f is of class C^r on $[0,1]$, $2 \leq r \leq +\infty$.

ii) The result of corollary A.10 is not true for arbitrary differentiability class, as shown by the following example.

Let f be an increasing diffeomorphism of class C^∞ of the interval $[0,1]$ onto itself which is a contraction on $[0,1)$; and let g denote an increasing diffeomorphism of class C^∞ of the interval $[f(\frac{1}{2}),\frac{1}{2}]$ onto itself which has contact of order ∞ with the identity map at $\frac{1}{2}$ and $f(\frac{1}{2})$. g can then be extended to an increasing homeomorphism of the interval $[0,1]$ onto itself, which is commuting with f and has an infinity of fixpoints, by letting $g(0) = 0$, $g(1) = 1$, and $g = f^n \circ g \circ f^{-n}$ on $[f^{n+1}(\frac{1}{2}), f^n(\frac{1}{2})]$.

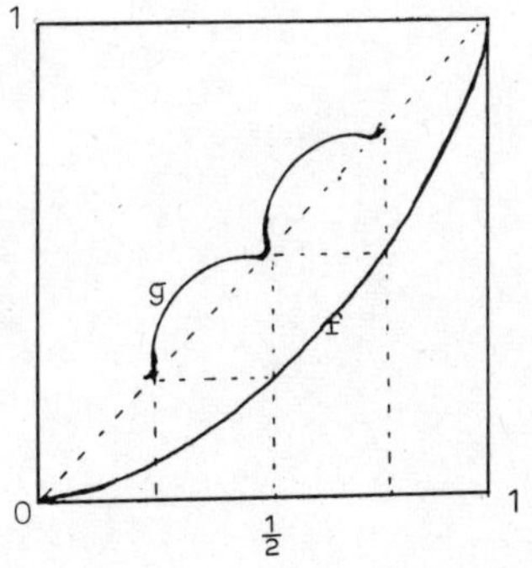

This homeomorphism g which is a diffeomorphism of class C^∞ on the open interval $(0,1)$ is not even of class C^1 on $[0,1]$.

iii) In the proof of theorem A.7 it was seen that a diffeomorphism g commuting with f has a derivative at 0 equalling 1. As the following exercise shows this phenomenon is quite general.

A.12. <u>Exercise</u>. Let f and g denote two local diffeomorphisms of class C^r, $2 \leq r \leq +\infty$, leaving the origin fixed and commuting. If f and g are different from the identity map near 0, and if f has contact of order k, $1 \leq k \leq r-1$, with the identity at 0, then the same holds for g.

Furthermore, for a finite k the map g is fully determined by its derivative $g^{(k+1)}(0)$.

Chapter III. Planar Vector Fields

The investigation of vector fields on open sets of the plane was developed by H. Poincaré and given a more precise form by I. Bendixson; it is nowadays known as the Poincaré-Bendixson theory. It is based in an essential way on the following version of Jordan's theorem (cf. [16]):

THEOREM. Let J denote a regular Jordan curve in the plane (i.e. the image of the circle S^1 by a differentiable embedding). The complement of J in $\mathbb{R}^2$ then has two components each of which has J as its boundary.

(From this version one can deduce Jordan's theorem for a piecewise regular Jordan curve J (having "corners") by observing that there is a continuous isotopy of the plane, which is smooth except for the corners and which transforms J into a regular Jordan curve.)

One of the components is bounded and is called the interior of J; the other is unbounded and is called the exterior of J.

1. LIMIT SETS IN THE PLANE

Let X be a vector field of class C^r, $1 \leq r \leq +\infty$ (respectively analytic) on an open and connected set U of the plane.

1.1. PROPOSITION. Every non-singular orbit of X is proper.

Proof. Let γ be a non-singular orbit of X. If γ is closed in the open set V of regular points of X, it is proper (I-2.6).

If γ is not closed in V it is intersected by a closed transversal T to X (proposition I-2.10). Along T the field X is then always directed either towards the interior, or towards the exterior of T. Hence the intersection of γ and T reduces to a single point. It follows from I-2.8 that γ is proper.

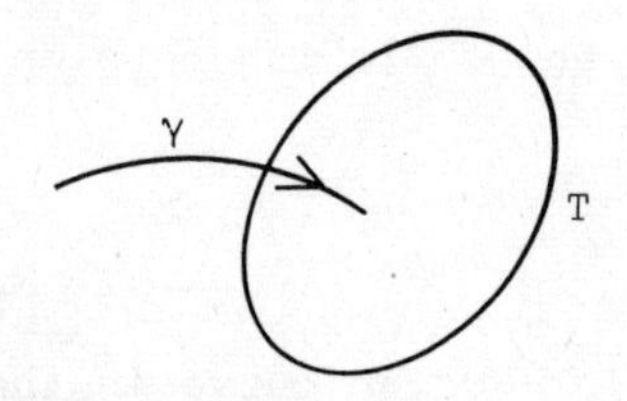

Q.E.D.

We deduce by I-3.6:

1.2. COROLLARY. Every minimal set of X is a singular point or a closed orbit.

1.3. Remark. The following two examples show that these results may fail for a non-orientable direction field. (The first example is due to Mme. Dubois-Violette, the second one to D. Epstein.)

i) Let us consider in the interior of the circle S^1 the family of circular arcs orthogonal to S^1 and with centres on a diameter D_1, and in the exterior of S^1 the family of circular arcs orthogonal to S^1 and having their centres on a different diameter D_2. We thus obtain on the open set of the plane complementary to the four points of intersection of D_1 and D_2 with S^1 a field E of directions whose orbits are formed by successive arcs of the first and second family.

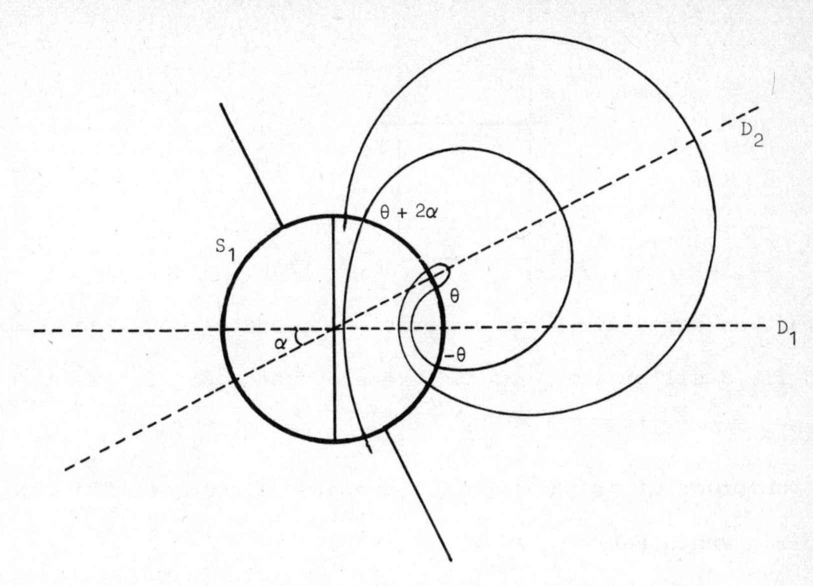

If the angle α of D_1 and D_2 is incommensurable with π every orbit of E is dense in U.

ii) Let G denote the transformation group of the plane generated by the translations $(x,y) \longmapsto (x+1,y)$ and $(x,y) \longmapsto (x,y+1)$, and by the symmetry $(x,y) \longmapsto (-x,-y)$. (By this group the torus T^2 appears as a two-sheeted covering space of the sphere S^2 with four ramification points which can be considered as the Riemann surface of an elliptic function.)

This group acts properly and freely on the open set

$$U = \mathbb{R}^2 - (\tfrac{1}{2}\mathbb{Z}) \times (\tfrac{1}{2}\mathbb{Z}).$$

The sketch below shows how we can identify the quotient space V of U by G with the sphere S^2 minus four points, or else with the plane $\mathbb{R}^2$ minus three points.

Every direction field on U which is invariant under G induces a direction field on V. In particular the direction field defined on U

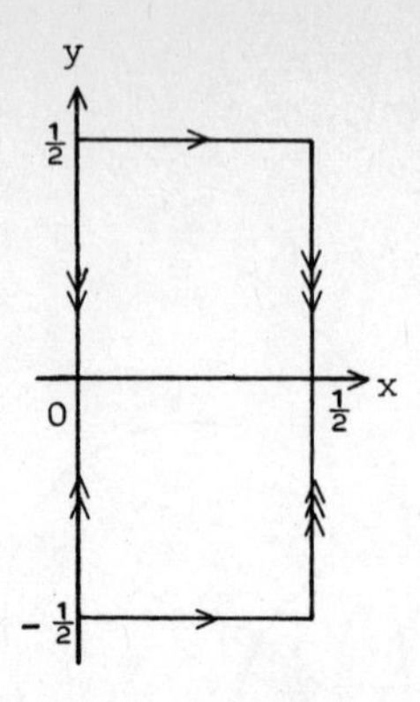

by the Pfaffian form $dy - \alpha dx$, α irrational, induces on V an analytic direction field all of whose orbits are everywhere dense (exercise ii) of II-1.12).

The proof of proposition 1.1 contains also a verification of the following statement:

1.4. PROPOSITION. For a vector field X on the plane $\mathbb{R}^2$ every non-closed orbit is either ω-stable or α-stable in Lagrange's sense.

1.5. PROPOSITION. Let γ be an orbit of X. If the field X has no singular point in U every orbit contained in the limit sets Ω_γ and A_γ is closed in U.

Proof. Let γ' be an orbit in Ω_γ. If it is not closed in U then it is intersected in a point u by a closed transversal T of X. If the field X is directed towards the interior of T then the positive half-orbit γ'^+_u lies in the interior of Γ, and the negative half-orbit γ'^-_u in the exterior.

Since γ' belongs to the closure of γ, the latter cuts T as well, in a point v, and the positive half-orbit γ_v^+ lies again in the interior of T. But then $\gamma_u'^-$ could not belong to Ω_γ contrary to the assumptions.

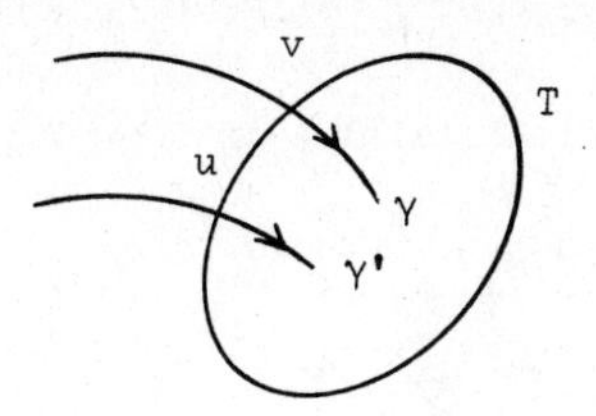

If the field X is directed towards the exterior of T we conclude as before by interchanging the rôles of the positive and negative half-orbits.
Q.E.D.

1.6 <u>COROLLARY</u> <u>(Theorem of Poincaré-Bendixson)</u>. Let γ be an orbit of X which in the sense of Lagrange is ω-stable (respectively α-stable). If the limit set Ω_γ (respectively A_γ) does not contain any singular point of X then it reduces to a periodic orbit of X.

In this case Ω_γ (respectively A_γ) is indeed compact and non-empty (proposition I-4.7), and every one of its orbits is periodic (proposition 1.5). But then it can contain only a single one (proposition II-7.11).

1.7. <u>COROLLARY</u>. Let γ be an orbit of X which is ω-stable (respectively α-stable) in Lagrange's sense. If the set Ω_γ (respectively A_γ) does not reduce to a periodic orbit then each of the limit sets of the orbits in Ω_γ (respectively A_γ) is a compact and connected non-empty set of singular points of X.

A non-singular orbit in Ω_γ is indeed (ω- and α-) stable in Lagrange's sense, and closed in the open set of regular points of X. Its limit sets are hence compact, connected, and non-empty, and consist of singular points of X belonging to Ω_γ.

If in particular the singular points of X are isolated then these limit sets reduce each to one singular point.

1.8. Exercises.

i) Let T denote a transverse arc to the field X. The successive intersections of an orbit γ of X with T form a monotonic sequence on T.

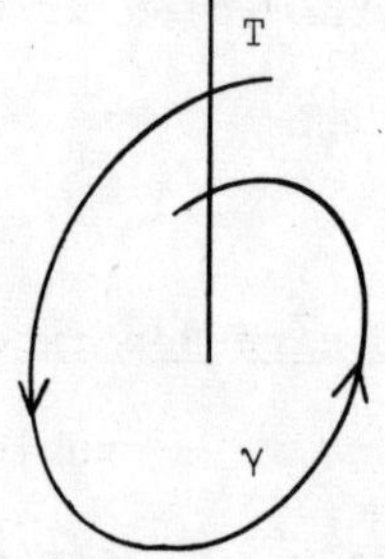

We deduce that the intersection of a limit set for X with a transverse arc (respectively a distinguished open set) contains at most one point (respectively one slice).

By the same technique it can be shown that a closed orbit of a planar vector field X cuts a transverse arc in at most one point.

ii) The set of non-singular orbits in a limit set is countable.

iii) Let γ denote an ω-stable orbit in Lagrange's sense whose ω-limit set contains a single singular point z of X. Then every orbit in Ω_γ has z as limit sets, and for every $d > 0$ the number of orbits in Ω_γ having a diameter greater than d is finite.

1.9. Remarks.

i) Let X be a vector field on $\mathbb{R}^2$. Then the field

$$Y: u \longmapsto \left[\exp(-\|X(u)\|^2 - \|u\|^2)\right] X(u)$$

has the same orbits as X, and can be extended as a vector field on the sphere S^2 vanishing at infinity.

This trick allows us, by considering the situation on the sphere S^2, to consider only (ω- and α-stable) orbits in Lagrange's sense, and to include in the above results the asymptotic behaviour of the unbounded orbits. An unbounded limit set, e.g., does not contain a compact component, etc.

ii) Let X denote a vector field in the plane $\mathbb{R}^2$ whose components P and Q are polynomials in x and y of degrees p and q respectively, $q \geq p$. According to Poincaré we can then use a finer compactification by considering $\mathbb{R}^2$ as the open set of finite points in the real projective plane $\mathbb{P}\mathbb{R}^2$.

The non-singular orbits of X are first of all interpreted as the orbits of the direction field defined on the open set of regular points of X by the Pfaffian form $\omega = Q\,dx - P\,dy$.

Let homogeneous coordinates (ξ, η, ζ) be introduced in $\mathbb{P}\mathbb{R}^2$, and let homogeneous polynomials be defined by

$$P^*(\xi,\eta,\zeta) = \zeta^p P(\frac{\xi}{\zeta},\frac{\eta}{\zeta}), \quad Q^*(\xi,\eta,\zeta) = \zeta^q Q(\frac{\xi}{\zeta},\frac{\eta}{\zeta}).$$

From ω we obtain, up to a factor ξ^{q+2}, the Pfaffian form on $\mathbb{R}^3$

$$\zeta Q^* d\xi - \zeta^{q-p+1} P^* d\eta + (\eta\zeta^{q-p} P^* - \xi Q^*) d\zeta ,$$

by the coordinate transformation $x = \frac{\xi}{\zeta}$, $y = \frac{\eta}{\zeta}$.

Because of its homogeneity this form defines a direction field on $\mathbb{P}\mathbb{R}^2$ outside of the set of finite singular points of X, and of the zeros of the polynomial $\eta P^*(\xi,\eta,0) - \xi Q^*(\xi,\eta,0)$ for $q = p$, or the zeros of the polynomial $\xi Q^*(\xi,\eta,0)$ for $q > p$ at infinity.

The orbits in the finite part of $\mathbb{P}\mathbb{R}^2$ of this direction field are the non-singular orbits of X, and the line at infinity is an invariant set of this direction field.

1.10. Exercises.

i) The above direction field is orientable for an odd q, and it is transversally orientable for an even q (cf. exercise I-5.11).

ii) By the above method we may study e.g. the behaviour at infinity of the orbits of the vector field $P = x^2 + y^2 - 1$, $Q = 5(xy - 1)$.

2. PERIODIC ORBITS

The theorem of Poincaré-Bendixson guarantees that a compact limit set contains a singular point or reduces to a periodic orbit.

In this section we shall be making some applications of this to the study of periodic orbits.

2.1. THEOREM. Every periodic orbit of the vector field X whose interior is contained in the domain of definition of X contains a singular point of X in its interior.

Proof. Let γ denote such a periodic orbit; then its interior is invariant under X and contains a singular point or a periodic orbit: For a non-singular and non-periodic orbit γ' lying in the interior of γ its limit sets indeed contain a singular point of a periodic orbit; but if one of them coincides with γ the two are disjoint.

Consider now such a periodic orbit γ whose interior does not contain a singular point of X. The set $\mathcal{P}$ of periodic orbits interior to γ and ordered by inclusion of interiors is inductive: if $(\gamma_i)_{i\in I}$ is a nested family of periodic orbits every orbit contained in the intersection of the compact sets bounded by these orbits γ_i has periodic orbits as limit sets. The set $\mathcal{P}$ has thus a minimal element; this contradicts, however, our initial remark above. Q.E.D.

2.2. <u>COROLLARY</u>. Every closed transversal of X whose interior is contained in the domain of X contains in its interior a singular point of X.

Assume e.g. that the field X is directed towards the interior along such a closed transversal Γ. Then the ω-limit set of an orbit intersecting T is in the interior of T and contains a singular point or a periodic orbit.

2.3. <u>COROLLARY</u>. A vector field defined on the plane $\mathbb{R}^2$ without singularities has neither a periodic orbit nor a closed transversal. Moreover every orbit of X is closed in $\mathbb{R}^2$.

Taking into account proposition 1.4 X would indeed have a periodic orbit if it had a non closed orbit.

The two following propositions are also immediate consequences of the theorem of Poincaré-Bendixson:

2.4. <u>PROPOSITION</u>. Let γ_1 and γ_2 denote two periodic orbits of X such that γ_1 lies in the interior of γ_2, and such that the open annulus C bounded by γ_1 and γ_2 belongs to the domain of X. (By Jordan's theorem the complement of $\gamma_1 \cup \gamma_2$ in $\mathbb{R}^2$ has three components: the interior of γ_1, the exterior of γ_2, and the "annulus" bounded by γ_1 and γ_2.) This annulus is then invariant under X. If it contains neither a singular point nor a periodic orbit of X one of the two following situations arises:

i) γ_1 is a limit cycle which is unstable in the exterior, and γ_2 is a limit cycle which is stable in tne interior;

ii) γ_1 is a limit cycle which is stable in the exterior, and γ_2 a limit cycle which is unstable in the interior.

In the first case every orbit in C has γ_1 as α-limit set and γ_2 as ω-limit set. In the second case the rôles of γ_1 and γ_2 are interchanged.

Of course, the transformation of X into -X interchanges these two situations.

2.5. PROPOSITION. Let T_1 and T_2 be two closed transversals of X such that T_1 lies in the interior of T_2, and such that the open annulus C bounded by T_1 and T_2 is contained in the open set of regular points of X. If then the field X points inward (respectively outward) on the boundary of $\bar{C}$, the annulus C contains a periodic orbit of X.

In view of theorem 2.1 such a periodic orbit is "between" T_1 and T_2 (i.e. it contains T_1 in its interior).

2.6. Examples.

i) The vector field with components $P = (\rho - 1)(3 - \rho)$ and $Q = 1$ on the half-plane $H = \{(\rho, \theta) \mid \rho > 0\}$ induces on the cylinder $\mathbb{R}^2 - \{0\}$ a vector field Y_1 without singularity which has as its only periodic orbits the circles C_1 and C_2 with centre 0 and radii 1 and 3. (They are unstable and stable limit cycles respectively.)

ii) Similarly the vector field with components $P = (\rho-1)(3-\rho)$ and $Q = \rho-2$ on H induces on $\mathbb{R}^2-\{0\}$ a vector field Y_2, without singularities as well, and having the same periodic orbits as Y_1 with the same stability properties.

These vector fields, however, are not topologically conjugate

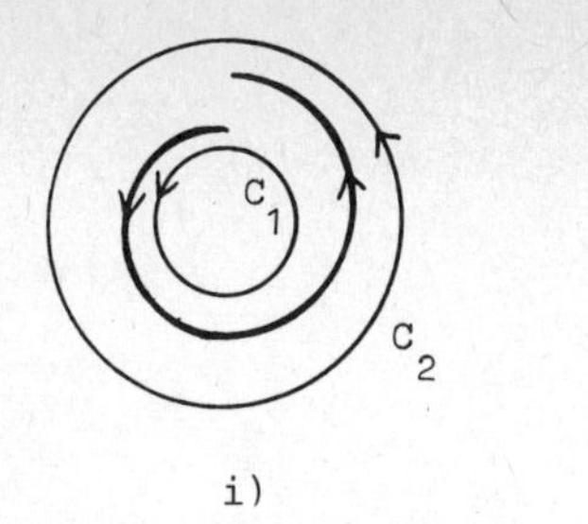

i)

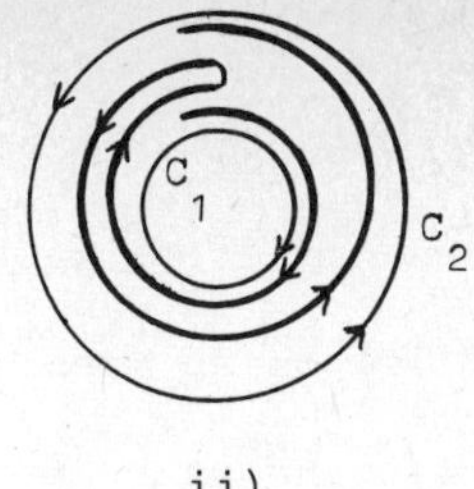

ii)

on the annulus $\overline{\Gamma} = \{(\rho,\theta) \mid 1 \leqslant \rho \leqslant 3\}$ (and neither are their associated direction fields conjugate): For the field Y_1 there are transverse arcs originating on C_1 and ending on C_2, which is not the case for Y_2.

2.7. Remarks.

i) The transformation $(\rho,\theta) \longmapsto (4-\rho,-\theta)$ (respectively $(\rho,\theta) \longmapsto (4-\rho,\theta)$) is an analytic conjugation of Y_1 to $-Y_1$ (resp. of Y_2 to $-Y_2$).

ii) Every direction field transverse to the vector field Y_2 is orientable (cf. proposition 6.2) and has a periodic orbit in the annulus Γ.

The following theorem emphasizes the importance of the two preceding examples [6]:

2.8. THEOREM. Under the assumptions of proposition 2.4 the vector field X on the closed annulus $\overline{C}$ is topologically conjugate to one of the two vector fields Y_1, Y_2 on the annulus $\overline{\Gamma}$.

If X is conjugate to Y_1, $\overline{C}$ is called a component of type I, and a component of type II if X is conjugate to Y_2. $\overline{C}$ is a component of type II if the orientations of γ_1 and γ_2 are compatible with an orientation of the manifold with boundary $\overline{C}$; otherwise it is a component of type I.

Proof. Let us consider situation i) of proposition 2.4, and assume that the orientations of γ_1 and γ_2 are not compatible with either of the two orientations of $\overline{C}$.

Choose a number ε between 0 and 1, and a conjugation h_1 of X to Y_1 (cf. II-7.10) on an open neighbourhood U_1 of γ_1 in $\overline{C}$ in such a way that the following statements hold:

i) $\overline{C} - U_1$ is a neighbourhood of γ_2 in $\overline{C}$;

ii) $U_1' = h_1(U_1)$ is the annulus defined by $1 \leqslant \rho < 1+\varepsilon$;

iii) h_1 is a diffeomorphism outside of γ_1.

Denote then by T_1' the circle $\rho = 1 + \frac{\varepsilon}{2}$ in $\overline{\Gamma}$, and by T_1 the closed transversal of X which is the inverse image of T_1' under h_1. (Note that T_1 intersects every orbit of X in C.)

Choose also in the interior of $\overline{C} - U_1$ a semi-open arc T which is transverse to X and whose endpoint u_o lies on γ_2; finally let T' be the segment defined in $\overline{\Gamma}$ by $\theta = 0$ and $3 - \varepsilon < \rho \leqslant 3$.

The Poincaré map P (respectively P') for γ_2 and the transverse arc T (respectively for C_2 and the transverse arc T') is a contraction. As the sketch below shows we can define a conjugation f from P to P' as follows:

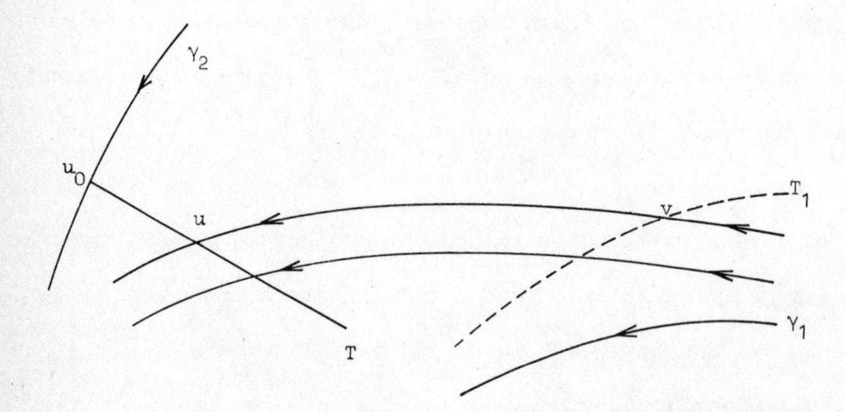

- $f(u_o)$ is the point $\theta = 0$, $\rho = 3$;

- for a point u in the interior of T the negative half-orbit γ_u^- intersects T_1 in point v after having cut n times the transversal T. The point f(u) is then the $(n+1)^{st}$ intersection of T' with the positive half-orbit $\gamma_{h_1(v)}^+$. (The continuity in u_o is guaranteed by our initial assumption.)

This conjugation is smooth except at u_o. It determines, as in II-7.3, a conjugation h_2 from X to Y_1 on an open neighbourhood U_2 of γ_2 in $\overline{C}$ which may be chosen to be disjoint with U_1 and such that $h_2(U_2)$ is the annulus defined by $3-\varepsilon < \rho \leqslant 3$.

The following essential property then holds: If u_1 in U_1 and u_2 in U_2 lie on a common orbit of X the points $h_1(u_1)$ and $h_2(u_2)$ are on a common orbit of Y_1.

As above denote by T_2' the circle $\rho = 3 - \frac{\varepsilon}{2}$ in $\overline{\Gamma}$, and by T_2 the closed transversal to X which is the inverse image of T_2' under h_2. (As before T_2 intersects also all orbits of X in C.)

Let Φ (respectively Ψ) be the flow of diffeomorphisms of $\overline{C}$ (respectively $\overline{\Gamma}$) generated by X (respectively Y_1). There exists then a smooth positive function α on T_1 (respectively a number $\tau > 0$) such that for every point u of T_1 (respectively u' of T_1') the intersection of the positive half-orbit γ_u^+ with T_2 (respectively $\gamma_{u'}^+$ with T_2') is the point $\Phi(\alpha(u),u)$ (respectively $\Psi(\tau,u')$).

Let us view now the annulus $\overline{C}$ as union of the following three compact sets bounded by γ_1, T_1, T_2, and γ_2:

$$K_1 = \gamma_1 \cup \Phi((-\infty,0] \times T_1),$$

$$K = \{\Phi(t\alpha(u),u) \mid u \in T_1 \text{ and } t \in [0,1]\},$$

$$K_2 = \Phi([0,+\infty) \times T_2) \cup \gamma_2 .$$

Finally define a conjugation k from X to Y_1 by setting:

$$k(u) = h_1(u) \quad \text{for } u \in K_1,$$

$$k[\Phi(t\alpha(u),u)] = \Psi(t\tau, h_1(u)) \quad \text{for } u \in T_1 \text{ and } t \in [0,1],$$

$$k(u) = h_2(u) \quad \text{for } u \in K_2 .$$

In the case where the orientations of γ_1 and γ_2 are compatible with an orientation of $\overline{C}$ the construction of the conjugation from X to Y_2 proceeds along analoguous lines. Q.E.D.

2.9 Remark. By suitably adapting the glueings along T_1 and T_2 we could make this conjugation smooth on C. We could even make it smooth on $\overline{C}$ by replacing in the examples 2.6 component $P = (\rho-1)(3-\rho)$ by a smooth function $f(\rho)$ which is positive on the interval $[1,3]$, vanishes at its endpoints, and is suitably chosen as a function of the vector field X.

The vector field Z with components $P = -y$ and $Q = x$ has no singularity on the cylinder $\mathbb{R}^2 - \{0\}$, and all its orbits are periodic.

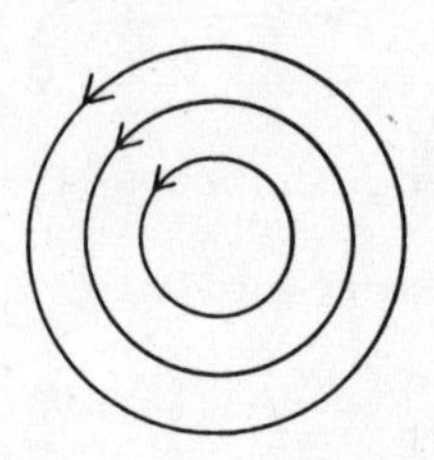

As before we denote now the annulus $1 \leqslant \rho \leqslant 3$ by $\overline{\Gamma}$, and we have then

2.10. PROPOSITION. Let γ_1 and γ_2 be two periodic orbits of the vector field X with γ_1 lying in the interior of γ_2, and such that the open annulus C bounded by γ_1 and γ_2 is contained in the open set of the regular points of X. If all orbits of X in C are periodic then the field X is differentiably conjugate on the annulus $\overline{C}$ to the field Z on the annulus $\overline{\Gamma}$.

If the Poincaré maps of γ_1 and γ_2 do not coincide with the identity map in the interior of γ_1, respectively in the exterior of γ_2, then the annulus $\bar{C}$ is called a <u>component of type III</u>.

<u>Proof</u>. Integrating e.g. a transverse vector field to X we can construct a transverse arc T to X in $\bar{C}$ with initial point on γ_1 and terminal point on γ_2, and intersecting every orbit in $\bar{C}$ in a unique point. The Poincaré map for γ_1 and this transverse arc is then defined on the whole of T and is the identity map. According to the method of proof of theorem II-7.3 it thus provides the desired conjugation. Q.E.D.

2.11. <u>Application: Vector field on an annulus without singularity</u>.

Let us look at the situation where γ_1 and γ_2 are two periodic orbits of X such that γ_1 lies in the interior of γ_2 and such that the open annulus C bounded by γ_1 and γ_2 is contained in the open set of regular points of X.

From theorem 2.1 it then follows that the periodic orbits of X in $\bar{C}$ are "nested", i.e. their set is totally ordered by inclusion of interiors; and the following properties hold:

i) The set of all periodic orbits is closed in $\bar{C}$;

ii) the set of components of type II in $\bar{C}$ is finite;

iii) if $\bar{C}$ does not contain any component of type II then the field X is differentiably conjugate on $\bar{C}$ to the suspension field of a diffeomorphism f of the segment $[0,1]$ leaving its endpoints fixed: indeed it is then possible to construct a transverse arc to X in $\bar{C}$ starting on γ_1 and ending on γ_2 (in this case the conjugation class of X is characterized by the fixpoint set of f and by the sign of $f(x) - x$ on each of the open intervals where f differs from the identity);

iv) if X is analytic either all its orbits are periodic, or $\bar{C}$

is a finite union of components of types I and II.

(From these properties we can deduce that the annulus $\overline{C}$ is diffeomorphic to the standard annulus $S^1 \times [0,1]$ which is a special case of the theorem of Schoenflies (cf. theorem 4.13).)

An analoguous investigation could be carried out in the case of two closed transversals T_1 and T_2 of X such that T_1 lies in the interior of T_2 and the annulus bounded by T_1 and T_2 is contained in the open set of regular points of X.

2.12. Exercise. Let $X = (P,Q)$ be a vector field on an open set U of the plane whose divergence $\frac{\partial P}{\partial x} + \frac{\partial Q}{\partial y}$ is strictly positive.

The field X then does not have a periodic orbit whose interior is contained in U.

It follows e.g. that the second order differential equation $x'' + f(x)\, x' + g(x) = 0$ has no periodic orbit in an interval where the sign of $f(x)$ does not change.

3. SINGULAR POINTS.

3.1. In this section we will consider a vector field $X = (P,Q)$ of class C^∞ (respectively analytic) on an open planar set having the origin as its only singular point.

An efficient way to study such a vector field near its singular point is to introduce polar coordinates.

For this let C denote the cylinder $\mathbb{R} \times S^1$, and let h be the map of C onto the plane defined by $x = \rho \cos \theta$ and $y = \rho \sin \theta$. X may then be viewed as image under h of the vector field Y defined in a neighbourhood of the base $\Gamma = \{0\} \times S^1$ of C, and with components

$$S = P(\rho\cos\theta,\rho\sin\theta)\cos\theta + Q(\rho\cos\theta,\rho\sin\theta)\sin\theta,$$

$$T = \frac{1}{\rho}\left[Q(\rho\cos\theta,\rho\sin\theta)\cos\theta - P(\rho\cos\theta,\rho\sin\theta)\sin\theta\right].$$

This vector field is of class C^∞ (respectively analytic) and has no singularity outside of Γ. The map h is an analytic conjugation from Y on $(0,+\infty)\times S^1$ (or $(-\infty,0)\times S^1$) to X on $\mathbb{R}^2 - \{0\}$.

3.2. The orbits of the fields X and Y are easy to describe if one of the functions $xP+yQ$, $xQ-yP$ identically vanishes.

In the following we make the assumption that these functions are not infinitely "flat" at the origin.

Then we may write: $\quad S(\rho,\theta) = \rho^m[S_m(\theta)+\sigma_m(\rho,\theta)],$

$$T(\rho,\theta) = \rho^{n-1}[T_n(\theta)+\tau_n(\rho,\theta)],$$

where S_m and T_n are non-vanishing homogeneous polynomials in $\cos\theta$ and $\sin\theta$ of respective degrees m+1 and n+1, and where σ_m and τ_n vanish for $\rho = 0$.

Hence we can replace Y by the vector field Z with components

$$S'(\rho,\theta) = S_m(\theta)+\sigma_m(\rho,\theta),$$

$$T'(\rho,\theta) = \rho^\mu[T_n(\theta)+\tau_n(\rho,\theta)],$$

for $n-1=m+\mu$, $\mu\geq 0$, and with components

$$S''(\rho,\theta) = \rho^\nu[S_m(\theta)+\sigma_m(\rho,\theta)],$$

$$T''(\rho,\theta) = T_n(\theta)+\tau_n(\rho,\theta),$$

for $m=n-1+\nu$, $\nu\geq 0$; outside of Γ these two fields have indeed the same behaviour.

In order to simplify our study <u>we will in addition assume that the functions $S_m(\theta)$ and $T_n(\theta)$ have no zeros in common</u>.

3.3. <u>Examples</u>. If the eigenvalues of X at the origin are both not zero we may reduce our investigation (after a linear transformation of variables if necessary) to one of the following three cases:

$$P = \lambda x + o(x,y) \quad \text{and} \quad Q = \mu y + o'(x,y),$$

$$P = \lambda x + y + o(x,y) \quad \text{and} \quad Q = \lambda y + o'(x,y),$$

$$P = \alpha x - \beta y + o(x,y) \quad \text{and} \quad Q = \beta x + \alpha y + o'(x,y),$$

where the functions o and o' are "flat" of order 1 at 0.

These cases yield respectively:

$$S_1 = \lambda \cos^2\theta + \mu \sin^2\theta \quad \text{and} \quad T_1 = (\mu-\lambda)\sin\theta\cos\theta,$$

$$S_1 = \lambda + \sin\theta\cos\theta \quad \text{and} \quad T_1 = -\sin^2\theta,$$

$$S_1 = \alpha \quad \text{and} \quad T_1 = \beta.$$

3.4. First case: $m \leqslant n-1$ and $S_m(\theta)$ never zero.

The field Z is then transverse to the base Γ of C, and hence the field X is, near 0, topologically conjugate to one of the linear vector fields $u \mapsto -u$, or $u \mapsto u$.

For this reason we call the origin a (stable or unstable) node of X (cf. II-3.2).

3.5. Second case: $m \leqslant n-1$ and $S_m(\theta)$ has zeros.

The number of these zeros is then at most 2(m+1). They occur in pairs $(\theta,\theta+\pi)$ and thus define at most m+1 straight lines passing through the origin. These are called the critical directions of X at 0.

The field Z is transverse to Γ in every point where the function S_m does not vanish. The orbit passing through such a point $(0,\theta)$ thus corresponds to an orbit of X having 0 as limit set and being tangent at 0 to the straight line with polar angle θ. This orbit and the one corresponding to $\theta+\pi$ determine a smooth arc passing through the origin.

In order to study the field Z near a critical direction Δ_o which corresponds to a zero θ_o of S_m, we choose θ_1 and θ_2 on either side of

θ_o and a cylindric strip on C

$$\Sigma = \{(\rho,\theta)\,|\,0 \leqslant \rho \leqslant \rho_1 \text{ and } \theta_1 \leqslant \theta_o \leqslant \theta_2\}$$

such that

i) the component T' of Z does not vanish on Σ for $\rho > 0$;

ii) the component S' of Z does not vanish on the vertical components $\Delta_1 = \{(\rho,\theta_1)\,|\,0 \leqslant \rho \leqslant \rho_1\}$ and $\Delta_2 = \{(\rho,\theta_2)\,|\,0 \leqslant \rho \leqslant \rho_1\}$ of the boundary of Σ, and vanishes only at the point $u_o = (0,\theta_o)$ on the circular component $J_o = \{(0,\theta)\,|\,\theta_1 \leqslant \theta \leqslant \theta_2\}$ of $\partial\Sigma$.

Every half-orbit contained in Σ then has the point u_o as limit set.

Assume first that θ_o is a zero of odd multiplicity p of S_m. By changing, if necessary, Z into -Z, we may restrict ourselves to the case where the derivative $S_m^{(p)}(\theta_o)$ is positive, and consider the following two possibilities:

i) $T_n(\theta_o) > 0$: the field Z then points inward on Δ_1 and on J_o for $\theta > \theta_o$, and outward on Δ_2 and on J_o for $\theta < \theta_o$. Hence there are two orbits γ_1 and γ_2 (respectively γ_1' and γ_2'), possibly coinciding, having u_o as ω-limit set (respectively α-limit set), and having the following properties:

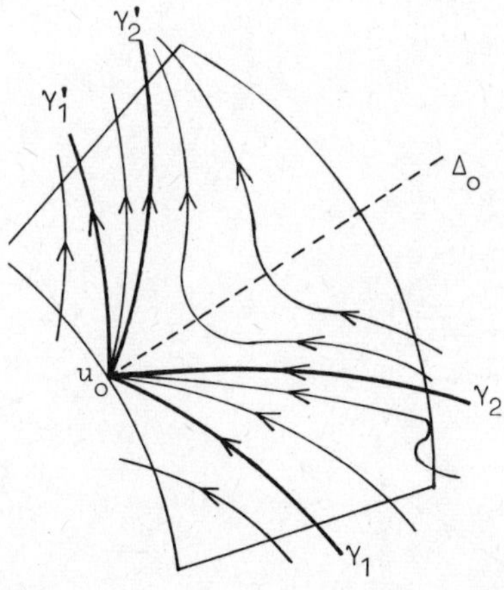

the orbits of Z below γ_1 or γ_1', or above γ_2 and γ_2' traverse Σ;

the orbits of Z lying between γ_1 and γ_2 (respectively γ_1' and γ_2') and sufficiently close to u_o have u_o as ω-limit set (respectively α-limit set);

ii) $T_n(\theta_o) < 0$: the field Z is then pointing outward on Δ_1 and on J_o for $\theta < \theta_o$, and inward on Δ_2 and on J_o for $\theta > \theta_o$. Thus there exists possibly an orbit γ (respectively γ') having u_o as ω-limit set (respectively α-limit set) and enjoying the following properties:

the orbits of Z above γ and γ' traverse Σ;

the orbits of Z below γ (respectively γ') and sufficiently close to u_o have u_o as ω-limit set (respectively α-limit set).

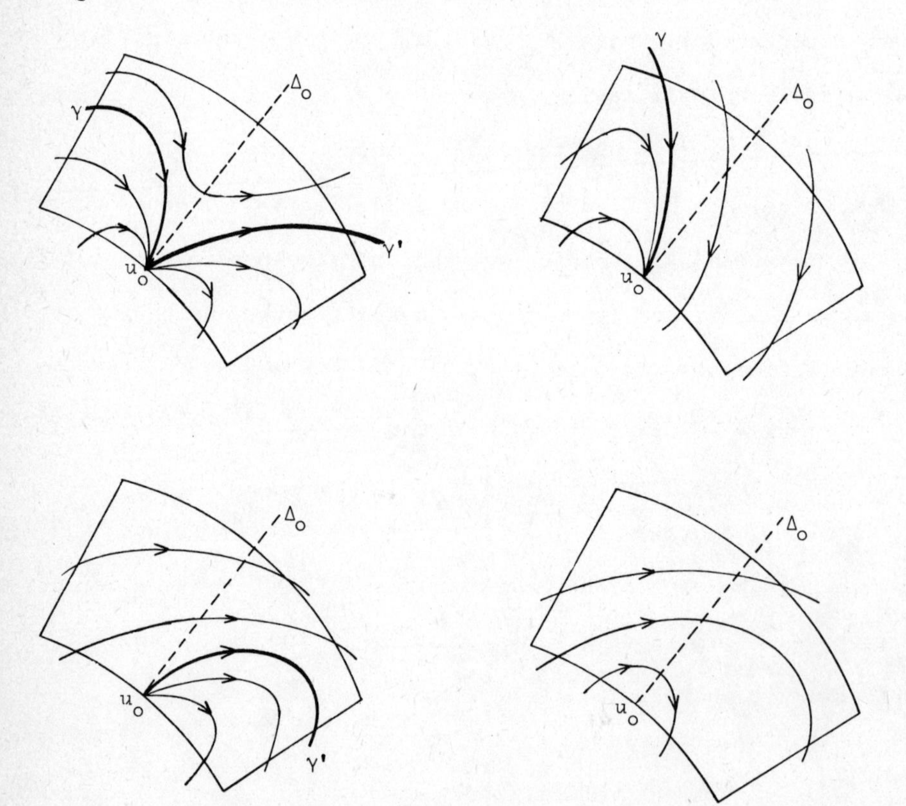

Assume now that θ_o is a zero of S_m of even multiplicity q. As before we may restrict ourselves to the case where the derivative $S_m^{(q)}(\theta_o)$ is positive, and where in addition $T_n(\theta_o) > 0$ (otherwise change θ into $2\theta_o - \theta$).

The field Z is then pointing inward on Δ_1 and on J_o, and outward on Δ_2. Hence there are two orbits γ_1 and γ_2 (possibly coinciding) having u_o as α-limit set, and perhaps an orbit γ having u_o as ω-limit set, with the following properties:

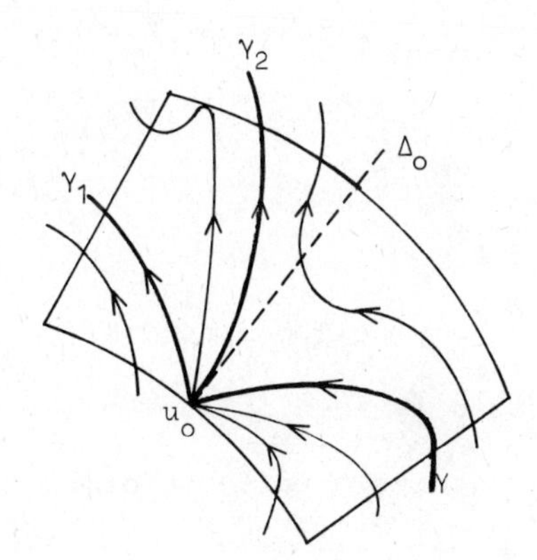

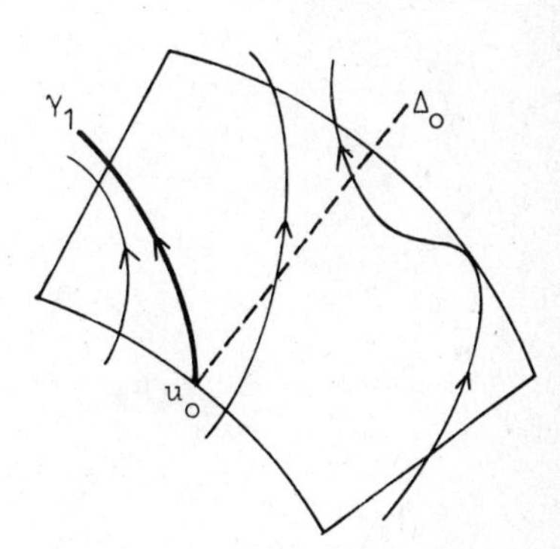

the orbits of Z below γ_1 and above γ and γ_2 traverse Σ,

the orbits of Z between γ_1 and γ_2, and sufficiently close to u_o have u_o as α-limit set;

the orbits of Z below γ have u_o as ω-limit set.

The separating orbits of Z occurring in these various cases correspond to orbits of X having 0 as limit set and being tangent at the origin to the appropriate critical direction.

Therefore if m is less than n-1 every orbit of X possessing C as limit set has a tangent at the origin, and every straight line passing through the origin is tangent to at least one orbit of this type, with the possible exception of certain critical directions.

3.6. <u>Third case: $m \geq n$ and $T_n(\theta)$ never vanishing.</u>

The base Γ of C is then a periodic orbit of Z, and there are essentially three possibilities (cf. II-7.10):

i) Γ is a stable (respectively unstable) limit cycle. Then the orbits of X close to the origin all tend toward 0 (respectively diverge away from 0) by spiralling: since this situation is conjugate to a similar linear case, 0 is called a (stable, resp. unstable) <u>focus</u>;

ii) all orbits of Z close to Γ are periodic. Then the same holds for all orbits of X close to the origin, and 0 is called a <u>centre</u>, as previously.

iii) Γ is not a limit cycle, and every neighbourhood of Γ contains a non-periodic orbit of Z. The origin is then neither a centre nor a focus, but every neighbourhood of 0 contains a periodic orbit of X.

In the first case the origin is a singular point for X which is asymptotically ω-stable (respectively α-stable) in Liapounov's sense. In the other two cases it is ω- and α-stable in Liapounov's sense, but there is no asymptotic stability.

If X is analytic the third case cannot occur. The problem of deciding whether the origin is a centre or a focus is then called <u>Poincaré's centre problem</u> [13]. In particular in example 3.3 we are confronted with this problem if the characteristic values of X at 0 are purely imaginary (cf. theorem II-6.4).

3.7. <u>Exercise</u>. Let X be a vector field in a neighbourhood of the origin whose components are of the form $P = y + o(x,y)$ and $Q = -x + o'(x,y)$ with o and o' flat of order 1 at 0.

If the direction field of X is symmetric with respect to the x-axis then the origin is a centre (e.g. for $P(x,-y) = -P(x,y)$ and $Q(x,-y) = Q(x,y)$).

This criterion applies in particular to the vector field corresponding to a second order differential equation of the type

$$x'' + f(x,x') + x = 0 .$$

3.8. Fourth case: $m \geq n$ and $T_n(\theta)$ having zeros.

The number of these zeros is then at most $2(n+1)$. They occur in pairs $(\theta, \theta+\pi)$ and thus define at most $n+1$ lines passing through the origin; these are called the <u>characteristic directions</u> of X at 0.

We note first the following properties:

i) the base Γ of C is invariant under Z;

ii) a non-singular orbit of X having 0 as limit set and having a tangent at the origin is tangent at 0 to one of the characteristic directions;

iii) a strip Σ which does not meet any characteristic direction does not contain in full any half-orbit, because Σ does not contain any singular point nor any periodic orbit of Z.

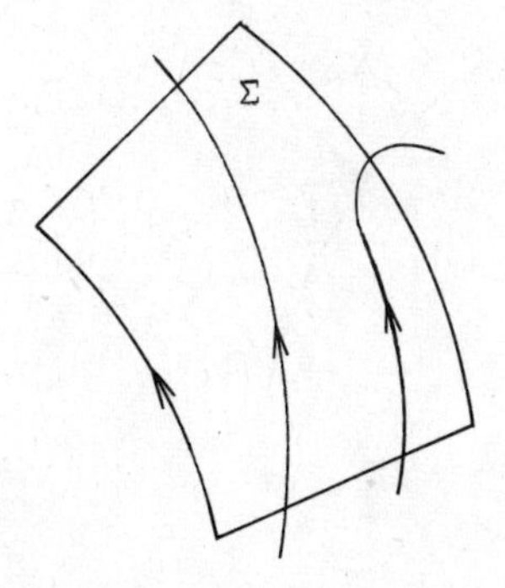

In order to study the field Z in the neighbourhood of a characteristic direction corresponding to a zero θ_o of T_n we choose θ_1 and θ_2 on either side of θ_o, and a strip

$$\Sigma = \{(\rho,\theta) \mid 0 \leqslant \rho \leqslant \rho_1 \text{ and } \theta_1 \leqslant \theta \leqslant \theta_2\}$$

such that

i) the component S" of Z does not vanish on Σ for $\rho > 0$;

ii) the component T" of Z does not vanish on the radial components Δ_1 and Δ_2 of the boundary of Σ, and does vanish only at the point $u_o = (0,\theta_o)$ on the circular component J_o of $\partial\Sigma$.

Every half-orbit contained in Σ has then the point u_o as limit set.

Assume first that θ_o is a zero of odd order p of T_n. Just as in 3.5 we may restrict ourselves to the case where the derivative $T_n^{(p)}(\theta_o)$ is positive. Then we may distinguish the two following possibilities:

i) $S_m(\theta_o) > 0$: the field Z then points outward on the boundary of Σ outside of J_o, and hence every orbit in Σ has the point u_o as α-limit set;

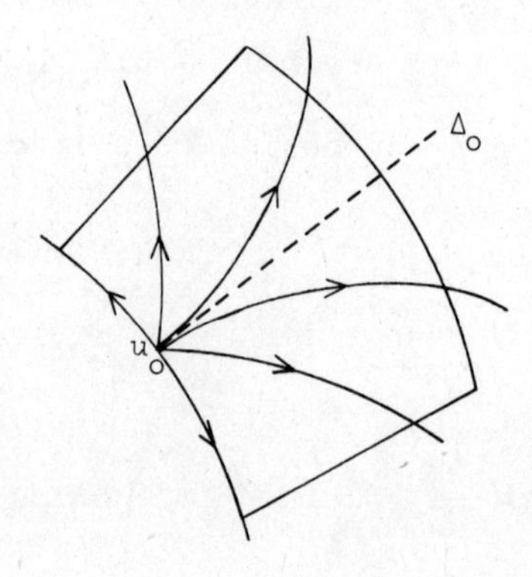

ii) $S_m(\theta_o) < 0$: the field Z then points inward on the circular component $J_1 = \{(\rho_1, \theta) \mid \theta_1 \leq \theta \leq \theta_2\}$ of the boundary of Σ, and outward on the radial components Δ_1 and Δ_2. Hence there exist two orbits γ_1 and γ_2 (possibly coinciding) which intersect J_1, have u_o as ω-limit set, and have the following additional properties:

- the orbits of Z lying between γ_1 and γ_2 have the point u_o as ω-limit;

- the orbits of Z to the left of γ_2 or to the right of γ_1 in $\Sigma - J_o$ traverse Σ.

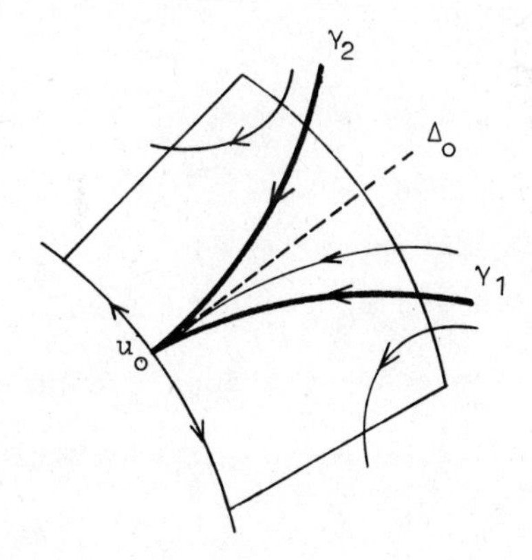

Assume now that θ_o is a zero of even order q of T . Once again we can restrict ourselves to the case where the derivative $T_n^{(q)}(\theta_o)$ is positive, and where in addition $S_n(\theta_o)$ is positive.

(See figure on next page.)

The field Z then points inward on Δ_1 and outward on Δ_2 and J_1. Hence there is an orbit γ intersecting $\Delta_2 \cup J_1$ (possibly lying on Γ) and having u_o as α-limit set, with the following additional properties:

- the orbits of Z lying on the left of γ have the point u_o as

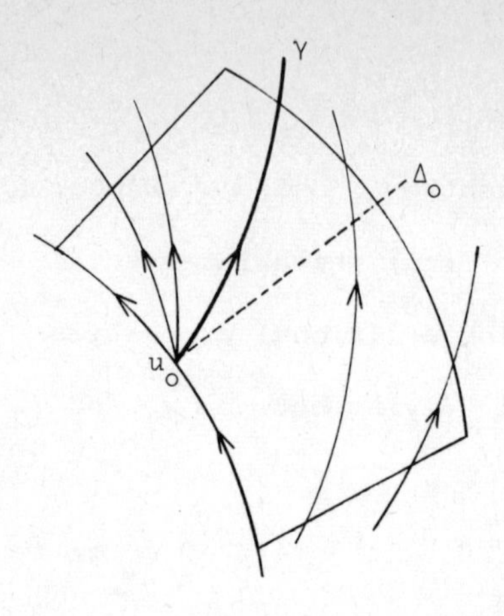

α-limit set;

— the orbits of Z on the right of γ traverse Σ.

In this fourth case every orbit having 0 as limit set is thus tangent at the origin to one of the characteristic directions.

3.9. Remark. Let $m \geq n$ and θ_o a simple zero of T_n with $T_n'(\theta_o) > 0$ and $S_m(\theta_o) < 0$. Then the above descriptions can be sharpened by observing that the orbits γ_1 and γ_2 coincide: the orbits tending toward u_o may then be parametrized in the form $\theta = \theta(\rho)$. Let then $\theta_1 = \theta_1(\rho)$ and $\theta_2 = \theta_2(\rho)$, $\theta_1 < \theta_2$, be two such orbits. For sufficiently small ρ we have the following estimates

$$\frac{d(\theta_2 - \theta_1)}{d\rho} < k\,\frac{\theta_2 - \theta_1}{\rho^{\nu}}\,, \qquad \frac{T_n'(\theta_o)}{S_m(\theta_o)} < k < 0\,.$$

This leads to a contradiction because of $\lim\limits_{\rho \to 0}(\theta_2 - \theta_1) = +\infty$.

3.10. Conclusions. Under the assumptions of section 3.2 we may give a global description of the behaviour of X near 0 by restating the preceding results as follows. (It is sufficient to restrict ourselves to the case where the origin is neither a focus nor a node, and has a neighbourhood which does not contain any periodic orbit of X.)

There exists then a finite collection $\gamma_o, \gamma_1, \ldots, \gamma_r$, $\gamma_r = \gamma_o$ of half-orbits of X, having the origin as limit set, and cyclically ordered by their behaviour at 0: for two consecutive orbits γ_i and γ_{i+1} one of the following statements holds:

a) the point 0 is the α-limit (respectively ω-limit) set for both γ_i and γ_{i+1}. All other orbits of X lying between γ_i and γ_{i+1} and sufficiently close to 0 have the origin as α-limit (respectively ω-limit) set as well;

b) the point 0 is the α-limit (respectively ω-limit) set for γ_i and the ω-limit (respectively α-limit) set for γ_{i+1}, and one of the following two possibilities occurs:

i) all orbits of X lying between γ_i and γ_{i+1} and sufficiently close to 0 have the origin as α-limit and ω-limit sets;

ii) no orbit of X between γ_i and γ_{i+1} and close to 0 remains near the origin.

The two orbits γ_i and γ_{i+1} are then said to bound

a <u>parabolic sector</u> in case a),

an <u>elliptic sector</u> in case bi),

a <u>hyperbolic sector</u> in case bii).

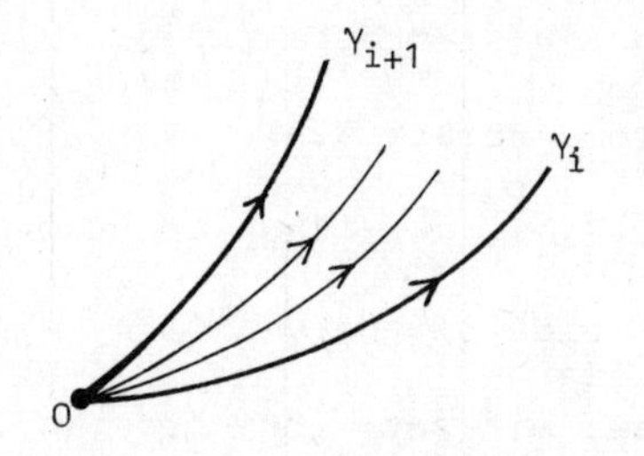

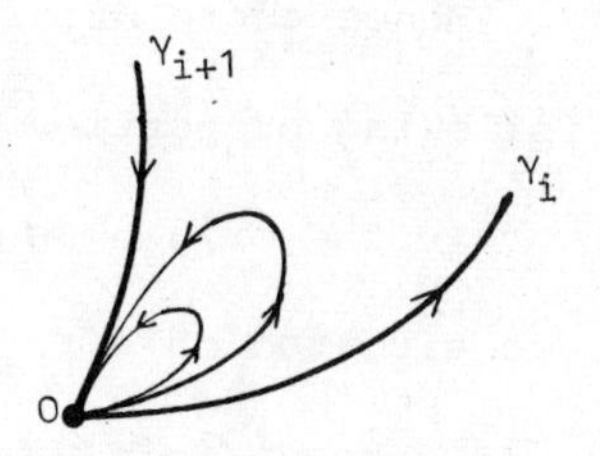

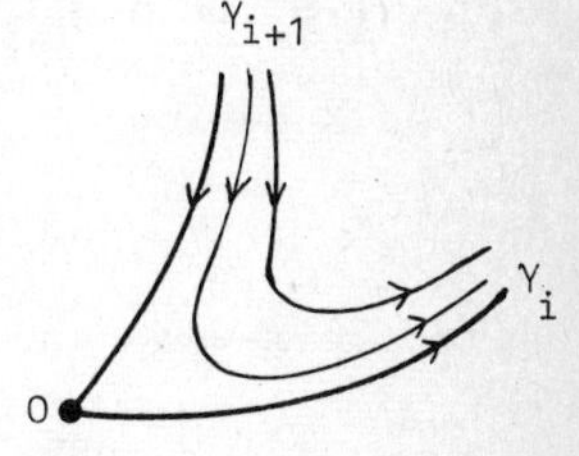

3.11. Exercise. Exhibit examples of vector fields having the origin as an isolated singular point and illustrating the various types of behaviour indicated above. If e.g. the components P and Q of X have the form $Ax - By$ and $Bx + Ay$, then $xP + yQ = (x^2+y^2)A$ and $xQ - yP = (x^2+y^2)B$.

We finish this section by a result on conjugation which completes theorem II-6.4 for a hyperbolic singular point in the plane.

3.12. THEOREM. Let u be a singular point of the planar vector field X, and let the eigenvalues of X at u be real, different from 0, and of opposite signs. Then the field X near 0 is topologically conjugate to the linear vector field $Y: (x,y) \mapsto (-x,y)$. (Hence u is then called a saddle.)

Proof. Assume $u = 0$, and the components of X of the form $\lambda x + o(x,y)$ and $\mu y + o'(x,y)$, $\lambda < 0 < \mu$, where o and o' are flat of order 1 at 0.

With the preceding notations the field Z has then the following components

$$S = \rho\left[\lambda \cos^2\theta + \mu \sin^2\theta + \sigma_1(\rho,\theta)\right],$$

$$T = (\mu - \lambda)\sin\theta \cos\theta + \tau_1(\rho,\theta) .$$

Here the characteristic directions of 3.8 are the two axes $x = 0$ and $y = 0$. They correspond to simple zeros of $T_1(\theta)$ where the product $T_1' \cdot S_1$ is negative. Hence there are four half-orbits γ_1, γ_2, γ_3, and γ_4 of X having the following properties (cf. remark 3.9):

i) γ_1 and γ_3 have the origin as ω-limit set; they are tangent to the x-axis and are situated on either side of the y-axis;

ii) γ_2 and γ_4 have the origin as α-limit set; they are tangent to the y-axis and lie on either side of the x-axis;

iii) the four sectors bounded by these half-orbits are hyperbolic.

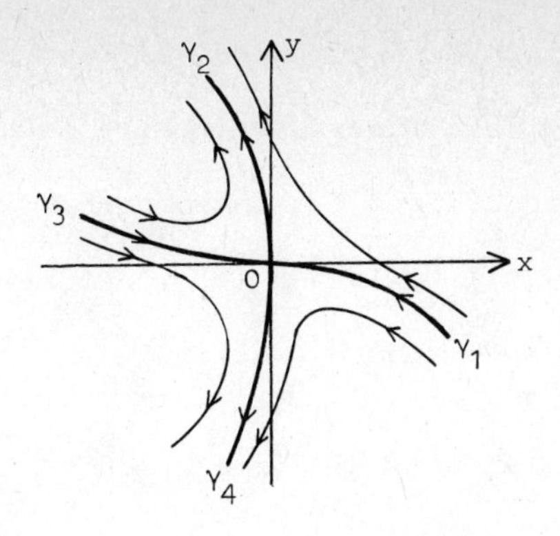

Choose then a compact set K bounded by

an arc of γ_1,

a segment Δ of the line $x = y$ which is transverse to X,

an arc J transverse to X having its initial point u on γ_1,

an arc of an orbit of X joining J to Δ.

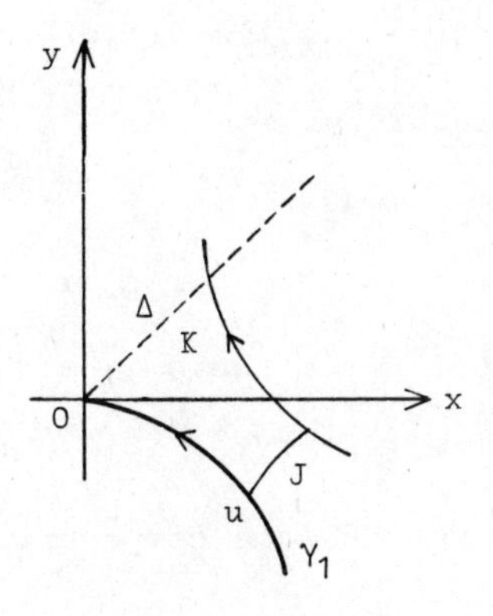

Let Φ denote the maximal flow generated by X. There is a smooth map α of $J - \{u\}$ into $[0,+\infty)$ such that $\Phi(\alpha(v),v)$ is for every $v \in J$ the intersection of the half-orbit γ_v^+ and Δ. This allows us to consider $K - \{0\}$ as the homeomorphic image under Φ of the subspace A of $[0,+\infty) \times J$ which is the union of $[0,+\infty) \times \{u\}$ and of the set $\{(t,v) \mid v \neq u,\ t \in [0,\alpha(v)]\}$.

Let us introduce now analoguous elements for the vector field Y, denoted by primed letters, and let j be a diffeomorphism of J onto J' carrying u into u', and h a homeomorphism of A onto A' of the form $(t,v) \longmapsto (h_v(t), j(v))$, with $h_v(t) = t$ for $v = u$, and for $v \neq u$ and $t \leq \beta(v)$. Here β is a smooth function satisfying

$$\frac{1}{3}\inf(\alpha,\alpha') \leq \beta \leq \tfrac{1}{2}\inf(\alpha,\alpha') \quad \text{on} \quad J - \{u\}.$$

We then obtain a conjugation k from X on K to Y on K', by defining $k(0)=0$ and $k(\Phi(t,v)) = \Phi'(h(t,v))$ for $(t,v) \in A$. The required conjugation is the result of glueing together such partial conjugations on either side of each of the four half-orbits γ_1, γ_2, γ_3, and γ_4. Q.E.D.

3.13. <u>COROLLARY</u>. Let u be a saddle point of X, and γ an orbit of X whose α-limit set contains u, but is different from $\{u\}$. Then the ω-limit set of γ does not contain u.

Indeed there is a transverse arc to X intersecting γ in at least two distinct points. Hence there is a closed transversal T of X intersecting γ (cf. remark ii) of I-2.11). But then the limit sets A_γ and Ω_γ are not situated in the same component of the complement of T.

3.14. <u>Remarks</u>.

i) Let u be a saddle of X. Then there are two half-orbits having u as ω-limit (respectively α-limit) set and being tangent at u to the characteristic direction corresponding to the negative (respectively positive) characteristic value of X at u. Their union together with the point u forms a smooth curve which is called the <u>stable</u> (respectively <u>unstable</u>) <u>manifold</u> of X at u. It consists of the points having u as ω-limit (respectively α-limit) set.

ii) If γ is an orbit of X having the saddle u as α- and ω-limit set, then the curve J formed by γ and u is a Jordan curve which is smooth except at u.

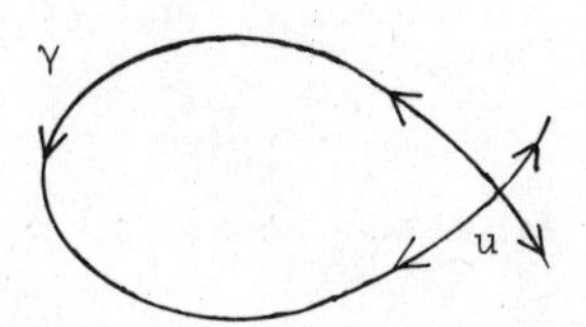

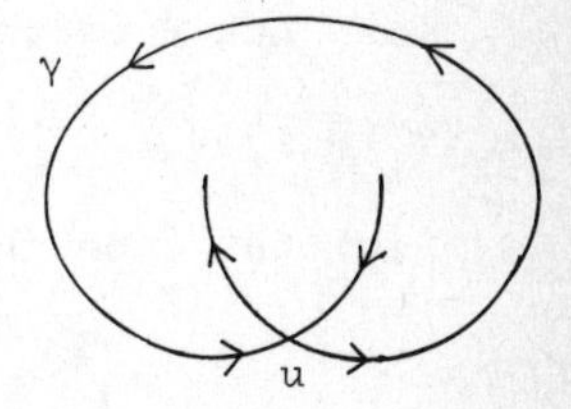

As in theorem 2.1 one confirms that the interior of J contains a singular point provided it is contained in the domain of definition of X.

4. THE POINCARE INDEX

Let X be a continuous vector field on a subspace A of the plane.

4.1. DEFINITION. Denote by $\gamma: S^1 \longrightarrow \mathbb{R}^2$ a loop whose image lies in the set of regular points of X. Then the degree $i_X(\gamma)$ of the map

$$\tilde{\gamma}: z \longmapsto X(\gamma(z))/\|X(\gamma(z))\|$$

of S^1 into itself is called the <u>Poincaré index</u> of X along γ (cf. [16]).

This index is thus an integer depending only on the oriented direction of X on the image of γ. It represents "the algebraic number of turns" which this direction makes when traversing γ in the oriented plane.

4.2. Elementary Properties.

i) If γ is a constant loop, or if X has constant direction on the image of γ, then $i_X(\gamma) = 0$.

ii) Let Y be obtained from X by a rotation of constant angle (e.g. $Y = -X$). Then $i_Y(\gamma) = i_X(\gamma)$.

iii) Let γ be the composition of two loops γ_1 and γ_2. Then $i_X(\gamma) = i_X(\gamma_1) + i_X(\gamma_2)$.

iv) Let $\bar{\gamma}$ be the loop $z \mapsto \gamma(\bar{z})$ (inverse to γ), then $i_X(\bar{\gamma}) = -i_X(\gamma)$.

v) Let γ_o and γ_1 denote two loops which are homotopic in the set B of regular points of X. Then $i_X(\gamma_o) = i_X(\gamma_1)$.

In particular the index of X along γ is invariant under parameter changes which preserve the orientation of S^1. It is thus possible to speak of the index of X along an oriented Jordan curve J without mentioning a parametrisation of J.

vi) Let X and Y be two continuous and nowhere vanishing vector fields on the image Γ of γ, and let X and Y be homotopic via non-vanishing vector fields. Then $i_X(\gamma) = i_Y(\gamma)$.

In particular:

Let X and Y be two continuous and nowhere vanishing vector fields on Γ which never point in opposite directions (i.e. $X/\|X\| + Y/\|Y\|$ never vanishes on Γ). Then $i_X(\gamma) = i_Y(\gamma)$.

vii) Let X be a vector field with smooth components P and Q, and let γ be a differentiable loop. Then

$$i_X(\gamma) = \frac{1}{2\pi}\int_{S^1} \tilde{\gamma}^*(d\theta) = \frac{1}{2\pi}\int_{S^1} \gamma^*\left(\frac{PdQ - QdP}{P^2 + Q^2}\right)$$

(Formula of Poincaré).

4.3. Examples. In this section γ denotes the canonical injection of S^1 into the plane.

i) Let X_o be the linear vector field $u \mapsto u$. Then $i_{X_o}(\gamma) = +1$.

ii) Let X be a linear vector field with components $P = \lambda x$ and $Q = \mu y$, $\lambda\mu > 0$. Then $i_X(\gamma) = +1$: if λ and μ are positive (respectively negative) then X and X_o (respectively X and $-X_o$) are never opposed to each other on S^1.

iii) Let Y be a linear vector field with components $P = \alpha x - \beta y$ and $Q = \beta x + \alpha y$. Then $i_Y(\gamma) = +1$: for $\beta \neq 0$ Y and X_o are indeed never in opposition on S^1.

iv) Let Z be a linear vector field with components $P = \lambda x$ and $Q = \mu y$, $\lambda\mu < 0$. Then by Poincaré's formula $i_Z(\gamma) = -1$.

4.4. Exercises.

i) Let X and Y denote two continuous unit vector fields on the image Γ of a loop γ, and let the indices of X and Y be different. Then there exist two points u and v on Γ such that $X(u) = Y(u)$ and $X(v) = -Y(v)$.

ii) Let P and Q denote the components of the vector field X, and let the zeros of $Q \circ \gamma$ on S^1 be isolated. Then $i_X(\gamma) = p - q$, where p (respectively q) is the number of those zeros in whose neighbourhoods P is positive and $Q \circ \gamma$ is increasing (respectively decreasing).

Let in particular γ be as in 4.3, and let X denote a linear hyperbolic vector field. Then $i_X(\gamma) = -1$ if the origin is a saddle for

X, and $i_X(\gamma) = +1$ otherwise.

iii) Let again γ be as in 4.3, and let X satisfy $X(-u) = X(u)$ (respectively $X(-u) = -X(u)$). Then the index of X along γ is an even (respectively odd) integer.

4.5. THEOREM. Let γ be a periodic orbit of the vector field X. Then $i_X(\gamma) = +1$.

Proof. Assume X to have norm 1 and γ to have period τ. Then choose a parametrisation of γ by an integral curve c of X defined on the interval $|0,\tau|$, and such that $c(0) = c(\tau)$ is a strict minimum of the ordinates on γ.

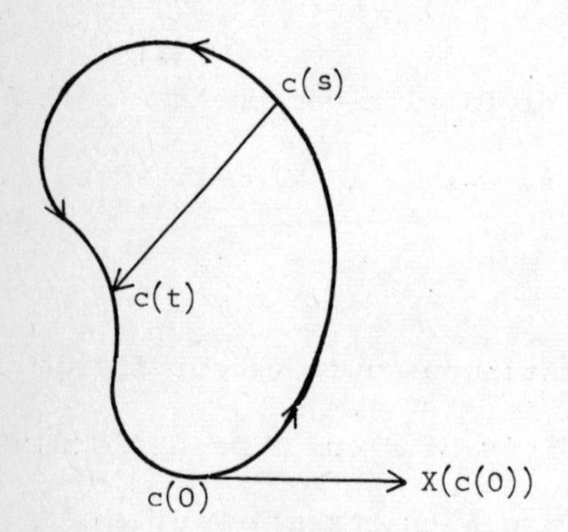

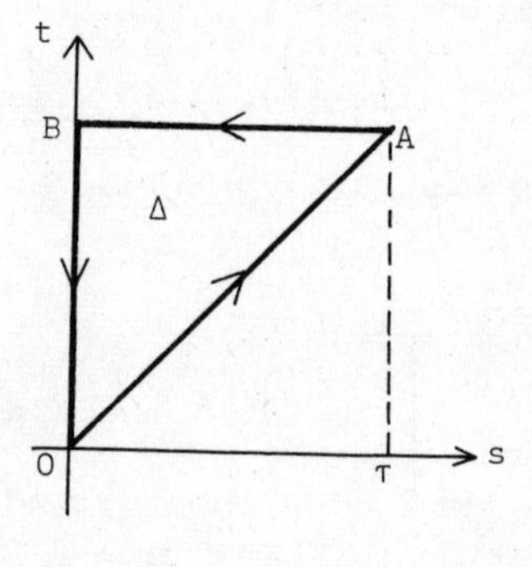

Let Δ denote the triangle bounded by the lines $s = 0$, $t = \tau$, and $s = t$. Then define on Δ a vector field Y by letting

$$Y(s,t) = \frac{\overrightarrow{c(s)c(t)}}{\|\overrightarrow{c(s)c(t)}\|} \quad \text{for} \quad s < t \quad \text{and} \quad (s,t) \neq (0,\tau) ,$$

$$Y(0,\tau) = -X(c(0)) ,$$

$$Y(s,s) = X(c(s)) .$$

This field is continuous and does not vanish on Δ, and its index along the boundary of Δ is obviously zero (properties i) and v) of 4.2).

The variation of the angle of Y is $2\pi i_X(\gamma)$ from 0 to A, and $-\pi$ from A to B and from B to 0. Q.E.D.

4.6. COROLLARY. Let T denote a closed transversal to the vector field X. Then $i_X(T) = +1$.

4.7. THEOREM. Let J denote a Jordan curve in the plane which bounds a compact surface M, and let $X = (P,Q)$ be a smooth and non-vanishing vector field on M. Then $i_X(J) = 0$.

(We orient a Jordan curve J in the plane coherently with the orientation induced by the orientation of the plane on the compact surface bounded by J, i.e. by "exterior normals".)

Since indeed the Pfaffian form $\omega = \dfrac{PdQ - QdP}{P^2 + Q^2}$ is defined and closed on M we find

$$i_X(J) = \frac{1}{2\pi}\int_J \omega = \frac{1}{2\pi}\int_M d\omega = 0 .$$

Hence we found also another proof of theorem 2.1, viz.

4.8. COROLLARY. Let X be a smooth vector field on an open set U of the plane. Every periodic orbit of X, and every closed transversal to X whose interior is contained in U, contains in this interior a singular point of X.

We also deduce (cf. V-2):

4.9. COROLLARY (Theorem of Poincaré-Hopf). Every smooth vector field on the sphere S^2 has a singular point.

Proof. Assume by contradiction that there is a smooth and non-vanishing vector field X on the sphere $S^2 = \{(x,y,z) \in \mathbb{R}^3 \mid x^2+y^2+z^2 = 1\}$. Let n denote the point (0,0,1), h the stereographic projection of the open set $U = S^2 - \{n\}$ onto the plane, and $\hat{X}$ the vector field $h^T \circ X \circ h^{-1}$. We may assume $X(n) = \frac{\partial}{\partial x}$.

Since the vector field $\hat{X}$ has no singularity its index along any Jordan curve in the plane vanishes. If $a > 1$ is sufficiently large this index along the circle J of centre 0 and radius a equals the index of the field $\hat{Y} = h^T \circ Y \circ h^{-1}$, where Y is the vector field on S^2 with components $1 - x^2$, $-xy$, and $-xz$ (cf. property vi) of 4.2).

The components of $\hat{Y}$ are $P(u,v) = \frac{1}{2}(1 + v^2 - u^2)$ and $Q = -uv$. It is easy to see (cf. exercise ii) of 4.4) that the index of $\hat{Y}$ along J is +2.

This completes the proof by contradiction. Q.E.D.

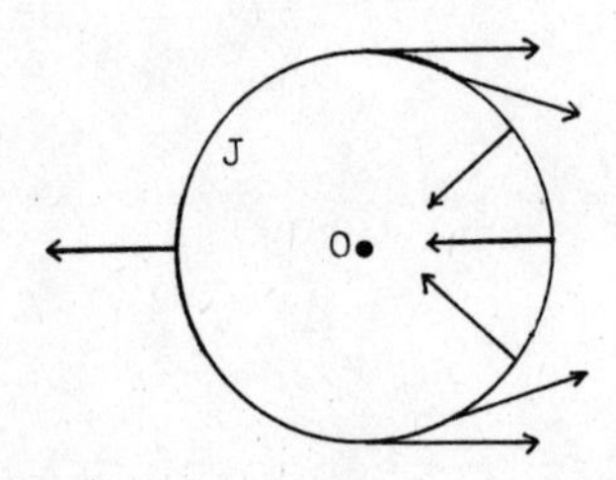

Alternately one might have argued that the unit tangent bundle of S^2 (cf. V-A.5) is diffeomorphic to the group $SO(3,\mathbb{R})$. If there were a non-vanishing vector field on S^2 then this unit tangent bundle would be diffeomorphic to the product $S^2 \times S^1$. This, however is impossible because of $\Pi_1(SO(3,\mathbb{R})) = \mathbb{Z}/2\mathbb{Z}$ and $\Pi_1(S^2 \times S^1) = \mathbb{Z}$.

The following proposition may be considered as a converse

of theorem 4.7:

4.10. PROPOSITION. Let J denote a planar Jordan curve bounding a compact surface M, and X a smooth and non-vanishing vector field on J whose index along J vanishes. Then there is an extension of X as a smooth and non-vanishing vector field on M.

Proof. Let h be a diffeomorphism of $J \times [0,+\infty)$ onto a collar U of J in M such that $h(u,0) = u$ for every $u \in J$ (cf. theorem I-1.23). Since the map $X/\|X\|$ defined on J is homotopic to a constant map because of the vanishing of the index, there is a smooth and non-vanishing extension of X on U such that $X(h(u,1)) = X(u)/\|X(u)\|$ and $X(h(u,t)) = \frac{\partial}{\partial x}$ for $t \geq 2$. X is then extended to M by letting $X = \frac{\partial}{\partial x}$ on $M - U$. Q.E.D.

The following theorem is an immediate generalisation of theorem 4.7 .

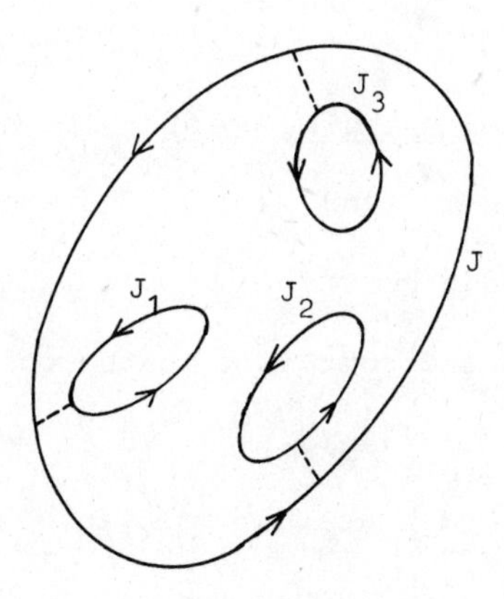

4.11. THEOREM. Let $J, J_1, \ldots, J_r$ denote $r+1$ disjoint Jordan curves in the plane with the following mutual positions:

$J_1,\ldots,J_r$ are contained in the interior of J,

the interiors of $J_1,\ldots,J_r$ are mutually disjoint.

Denote by M the compact and connected surface bounded by these curves in $\mathbb{R}^2$, and by X a smooth and non-vanishing vector field on M. Then

$$i_X(J) = \sum_{\alpha=1}^{r} i_X(J_\alpha).$$

4.12. Exercises.

i) The theorems 4.7, 4.11, and corollary 4.9 remain valid for merely continuous vector fields.

ii) Proposition 4.10 may be generalized in the situation of theorem 4.11. In particular we obtain the following results (cf. 2.11):

4.13. THEOREM (Schoenflies). The compact and connected surface bounded in the plane by a Jordan curve J (respectively by two nested Jordan curves J_1 and J_2) is diffeomorphic to the disc $\mathbb{D}^2$ (respectively to the annulus $S^1 \times [0,1]$).

4.14. Index of an isolated singular point. Let X be a smooth vector field on an open set U in $\mathbb{R}^2$, and $v \in U$ an isolated singular point of X. If J_1 and J_2 are two Jordan curves in U on which X does not vanish and whose interiors lie in U and contain v as the only singular point of X, then we have by theorem 4.11 $i_X(J_1) = i_X(J_2)$. Hence it is possible to define the index $i_X(v)$ of X at the isolated singular point v as the index of X along an arbitrary Jordan curve J in U with the above properties and oriented according to the convention of section 4.7.

4.15. Examples.

i) If every neighbourhood of v contains a periodic orbit of

X (in particular if v is a centre in the sense of section 3.6) then $i_X(v) = +1$.

ii) Let X be a linear hyperbolic vector field. Then $i_X(0) = -1$ for a saddle at 0, and $i_X(0) = +1$ otherwise (cf. exercise ii) of 4.4, or proposition 4.20).

4.16. Exercises.

i) Let $X = (P,Q)$ be a smooth vector field near the origin of $\mathbb{R}^2$ having 0 as an isolated singular point, and let P be an even and Q an odd function. Then $i_X(0) = 0$.

ii) Let p (respectively q) denote the number of elliptic (respectively hyperbolic) sectors of X at 0 in the situation of section 3.10. Then $i_X(0) = \frac{1}{2}(p-q+2)$. It follows that p and q have the same parity.

4.17. PROPOSITION. Let v be a non-degenerated singular point of a smooth vector field X on an open set of the plane. If the characteristic values of X at v are real and of different signs (i.e. if v is a saddle), then $i_X(v) = -1$, otherwise $i_X(v) = +1$.

Proof. We may assume $v = 0$ and write $X = Au + o(u)$, where A is an invertible matrix and o is flat of order 1 at 0. For sufficiently small $r > 0$ (viz. such that $\|o(u)\| < \|A\| \, \|u\|$ for every $u \neq 0$, $\|u\| < r$) the field X and the linear field $Y: v \longmapsto Av$ are never in opposition on the circle J of centre 0 and radius r. Hence $i_X(0) = i_Y(0)$, and we may apply example ii) of 4.15. Q.E.D.

4.18. THEOREM. Let J denote a Jordan curve in the plane bounding a com-

compact surface M, and X a smooth vector field on M, non-vanishing on J, and having in the interior of J a finite number of zeros $v_1,\dots,v_r$. Then

$$i_X(J) = \sum_{\alpha=1}^{r} i_X(v_\alpha).$$

Indeed we may surround every v_α by a Jordan curve J_α within J such that $J_1,\dots,J_r$ are disjoint and have mutually disjoint interiors. Then make use of theorem 4.11.

4.19. Exercise: roots of a polynomial. Let $P(z) = a_n z^n + \dots + a_o$ be a polynomial of degree n with complex coefficients, and let v denote a root of P.

i) Viewing v as an isolated zero of the vector field X on $\mathbb{R}^2$ corresponding to P the index of X at v becomes the multiplicity of v as root of P.

ii) The roots of a polynomial are continuous functions of the coefficients.

iii) If the coefficients $a_{n-r+1},\dots,a_n$ tend to 0, then r of the roots of P tend to infinity.

4.20. PROPOSITION (invariance of the index under diffeomorphisms). Let v be an isolated singular point of the smooth vector field X on an open set U of the plane, and let h be a diffeomorphism of U onto an open set V of $\mathbb{R}^2$; furthermore let $Y = h^T \circ X \circ h^{-1}$ (a vector field on V). Then

$$i_X(v) = i_Y(h(v)).$$

Proof. For a translation h this result is obvious; hence we may assume that $v = 0$ and $h(0) = 0$. We may also assume that h preserves the orientation of the plane, because the conclusion is equally evident for a symmetry $h:(x,y) \longmapsto (x,-y)$.

Let then denote by γ a circle of centre 0 and radius r lying in U, and such that there is no other singular point of X except 0 on the disc which it bounds. Let $\gamma' = h(\gamma)$.

Since h preserves the orientation the index of Y at 0 is the degree of the map $Y/\|Y\|$ on γ', or also of the map $h^T \circ X/\|h^T \circ X\|$ on γ. It thus equals the index of $h^T \circ X$ along γ. For sufficiently small r this is also the index of $h^T(0)X$ along γ. This finally agrees with the index of X at 0, because the group of linear automorphisms of the plane preserving the orientation is arcwise connected. * Q.E.D.

4.22. Index of a direction field. A direction field E on a subspace A of the plane is determined by a map h of A into the projective space of lines at 0 which can be identified with the circle S^1. This direction field is orientable if and only if this map can be factored via the two-sheeted covering map of S^1 corresponding to the projection $p:z \longmapsto z^2$ of S^1 onto itself, i.e. if there exists a map k of A into S^1 satisfying $h = p \circ k$ (cf. I-5).

The index $i_E(\gamma)$ of E along a loop $\gamma: S^1 \longrightarrow A$ is then defined as half of the degree of the map $h \circ \gamma$. This index is thus half of an integer; it is an integer if and only if E is orientable on the image of γ. Furthermore the indices of a non-vanishing vector field and of its direction field along the same loop coincide.

The results of 4.5 and 4.11 may be generalized to direction fields. In particular note the following consequence of the theorem of Schoenflies:

---

* Actually, the index is also invariant under homeomorphisms, as may be gathered from exercise ii) in 4.16.

4.22. PROPOSITION. Let J be a Jordan curve in the plane bounding a compact surface M. Then there is no direction field on M having J as a periodic orbit or as a closed transversal.

5. PLANAR DIRECTION FIELDS

This section is inspired by [4], and the dissertations of E. Fédida and Mme.M.P.Muller (Strasbourg 1973 and 1975).

5.1. Let E be a direction field in the plane. It may be considered as the direction field of a smooth, complete, and non-vanishing vector field X because it is orientable (cf. corollary I-5.8). It has neither a periodic orbit nor a closed transversal, and every orbit of E is closed in $\mathbb{R}^2$ (cf. corollary 2.3). Hence the following statements hold:

i) the complement of an orbit γ of E in the plane has two components (otherwise γ would intersect a Jordan curve in a single point and could not be closed);

ii) a transverse arc J of E cuts every orbit of E in at most one point;

iii) if J_1 and J_2 are two open arcs which are transverse to E then the union of the orbits of E meeting both J_1 and J_2 is a connected open set;

iv) if J is a transverse arc to E the flow Φ generated by X induces a diffeomorphism of $\mathbb{R} \times J$ onto an open set which is invariant for E.

Hence we conclude:

5.2. PROPOSITION. The space V of orbits of E is a connected one-dimensional manifold with a countable base. The complement of each of its points is an open set having two components. The projection p of $\mathbb{R}^2$ onto V is a locally trivial fibration into straight lines.

We impose on the projection p the condition of being a diffeomorphism for each transverse arc to E and thus provide the above manifold V with a differentiable structure, of the same differentiability class as E.

5.3. Remarks.

i) The examples below show that in general the manifold V is non Hausdorff. A point u of V which is not separated from another point v of V (i.e. such that every neighbourhood of u meets every neighbourhood of v) is called a branch point of V. The corresponding orbit $p^{-1}(u)$ is called separating, or a separatrix of E, and the orbits $p^{-1}(u)$ and $p^{-1}(v)$ are not separated.

The set of branch points of V is countable: in the situation iii) of 5.1 the arcs J_1 and J_2 determine at most two couples of non-separated points.

ii) If the manifold V is Hausdorff it is diffeomorphic to the line $\mathbb{R}$ (theorem I-5.10). Since every locally trivial bundle on $\mathbb{R}$ is trivial (cf. [18]), it follows that E is differentiably conjugate to the direction field whose orbits are the lines parallel to the x-axis.

5.4. Exercise: Simple branch spaces.

The simplest one-dimensional non-Hausdorff manifolds having the properties listed in proposition 5.2 are the simple branch spaces which are constructed as follows.

Let Σ denote the topological sum of two copies $\mathbb{R}_1$ and $\mathbb{R}_2$ of the real line $\mathbb{R}$, and let f be an increasing C^r-diffeomorphism of the interval $(0,+\infty)$. The space B_f obtained from Σ by identification of the points $x_1 \in \mathbb{R}_1$ and $x_2 \in \mathbb{R}_2$ for $x_1 > 0$, $x_2 > 0$ and $x_2 = f(x_1)$, is then a one-dimensional manifold with two branch points having the required properties.

Denoting by $\pi : \Sigma \longrightarrow B_f$ the projection, and by U_1 and U_2 the open sets $\pi(\mathbb{R}_1)$ and $\pi(\mathbb{R}_2)$, this manifold is provided with the C^r smooth structure determined by the two charts $(U_1, (\pi|\mathbb{R}_1)^{-1})$ and $(U_2, (\pi|\mathbb{R}_2)^{-1})$. The coordinate transformation $(\pi|\mathbb{R}_2)^{-1} \circ (\pi|\mathbb{R}_1)$ coincides with f. The following statements hold:

i) the two branch spaces B_f and B_g are C^s diffeomorphic, $0 \leqslant s \leqslant r$, if and only if there exist two increasing diffeomorphisms φ and Ψ of class C^s of $\mathbb{R}$ preserving the origin and such that $g \circ \varphi = \Psi \circ f$ or $g^{-1} \circ \varphi = \Psi \circ f$ on $(0,+\infty)$. In particular B_f and $B_{f^{-1}}$ are diffeomorphic.

ii) Every manifold B_f is contractible and homeomorphic to the simple branch space B corresponding to the identity map of $(0,+\infty)$.

iii) If $f = x^{\alpha}$ and $g = x^{\beta}$ the two branch spaces B_f and B_g are not C^1-diffeomorphic for $\alpha \neq \beta$. Contrary to the situation for one-dimensional Hausdorff manifolds, B is therefore a topological manifold having non-diffeomorphic smooth structures. (In the Hausdorff case the first and much more difficult examples of such a phenomenon were given by J.Milnor for the sphere S^7: Ann.of Math.64, 1956, pp.399-405.)

iv) For $f = x^n$, $n \in N$, every smooth function on B_f is flat of order n-1 at the branch point which is the image of the origin of $\mathbb{R}_1$ under π. It is thus possible to produce a one-dimensional manifold with the properties of 5.2 whose set of branch points is dense and which every smooth function is constant!

vi) Every one-dimensional manifold with the properties of 5.2 and having exactly two branch points is diffeomorphic to a simple branch space.

5.5. Exercise. Let V be the orbit space of a smooth direction field E in $\mathbb{R}^2$. The relation "u and v are non-separated points of V" is reflexive and symmetric, but in general not transitive. A branching class of V is then an equivalence class which does not reduce to a point with respect to the equivalence relation generated by the above relation.

i) A branching class C appears in a single one of the two components of the complement of every point of V-C.

ii) Let C and C' be two different branching classes. Then there is a point u outside of C and C' such that C and C' are in distinct components of $V - \{u\}$.

iii) Let λ denote a smooth loop in V. If the image of λ does not contain a branching point of V then λ' vanishes in at least two points. If the image of λ contains p branching points belonging to a common branching class then λ' vanishes in at least 2p points. If the image of λ contains q arbitrary branching points then λ' vanishes in at least q+2 points.

iv) Let E be defined by an algebraic Pfaffian form Pdx+Qdy, with P and Q polynomials of degree less than m. Then the space V has at most 2m branching points. (The points where the circle of centre 0 and of radius r touches E are the common zeros of the polynomials $x^2+y^2-r^2$ and $xQ - yP$.)

5.6. Examples.

i) The orbit space of a direction field E defined by a Pfaffian form of type $dy - a(x)dx$ is diffeomorphic to R. The same holds by exercise II-2.7 if E is defined by a Pfaffian form

$$dy - (a(x)y + b(x))dx .$$

ii) The orbit space of the direction field defined by the Pfaffian form $(\alpha - x)(1+x)dy - xdx$, $\alpha > 0$, is C^1-diffeomorphic to the simple branch space B_f for $f = x^{\alpha}$.

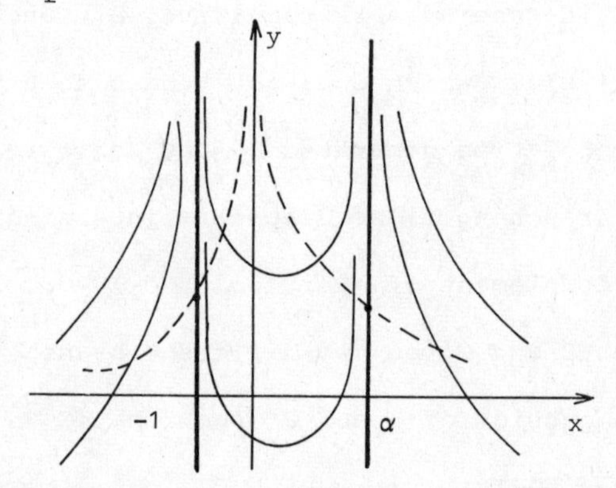

Note that contrary to the theorems of Feldbau (cf. [18]) in the Hausdorff case we obtain here a locally trivial fiber space with a contractible base which is not rivial, and a contractible fiber without having a section.

iii) By the change of variable $y = \tan u$ the direction field defined by a Pfaffian form of type $dy - a(x)(y^2+1)dx$ is reduced to the direction field defined on the strip $R \times (-\frac{1}{2}\pi, +\frac{1}{2}\pi)$ by the form $du - a(x)dx$. In particular we see that the orbit space defined by $dy - 2x(y^2+1)dx$ is the simple branch space B. Similar examples allow a great deal of variation in the examples of orbit spaces.

5.7. Remark. The orbit space is not sufficient to characterise the topological conjugation classes of diredtion fields in the plane, as shown by the following sketches.

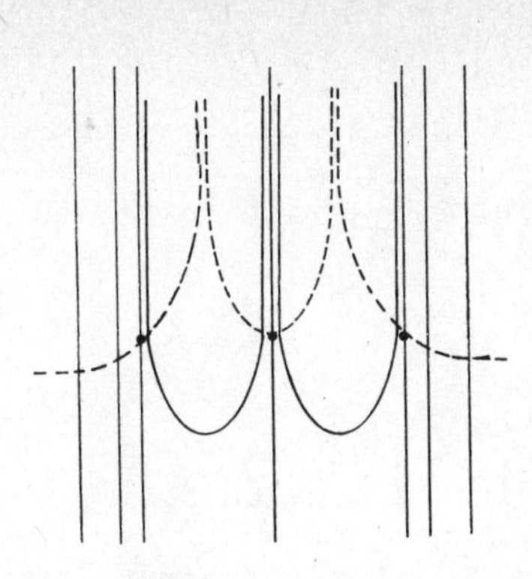

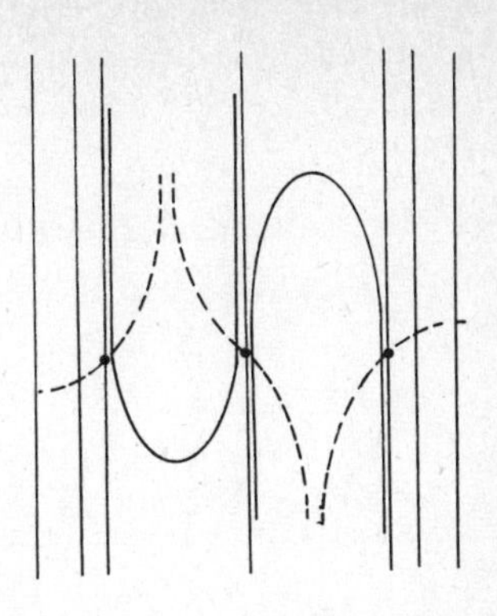

One can show, however, that all direction fields having a simple branch space as orbit space are topologically conjugate.

5.8. DEFINITION. Let E denote a direction field of class C^r on $\mathbb{R}^2$. A <u>(first) integral</u> of class C^s, $0 \leq s \leq r$, for E is a C^s-function on $\mathbb{R}^2$ which has no local extremum and which is constant on every orbit of E.

An integral of class C^s for E thus corresponds to a C^s-function on the space V of orbits of E which is a local diffeomorphism of V into $\mathbb{R}$.

5.9. <u>THEOREM</u> (Kaplan). Every planar direction field of class C^1 has a continuous first integral.

<u>Proof</u>. It will be sufficient to show that the orbit space V of a direction field E may be mapped into $\mathbb{R}$ by a local diffeomorphism.

Let $(U_p)_{p \in N}$ denote a covering of V by open sets which are diffeomorphic to R, and which are indexed in such a way that the union

$V = U_1 \cup \ldots \cup U_m$ is connected for every integer m. Let h_n be a local diffeomorphism of V_n into the interval $(-n,+n)$. It is monotonic on the intersection $I = V_n \cap U_{n+1}$. Since I is connected h_n can be extended continuously as a monotonic function h_{n+1} into U_{n+1} with values in $(-n-1,n+1)$.

Q.E.D.

5.10. Remarks.

i) Using the ideas of part iv) of section 5.4 we can construct a planar direction field of class C^∞ without a first integral of class C^1 (example of Wazewski).

For a direction field E of class C^r, $1 \leq r \leq +\infty$, however, the two following results hold:

If the set of separating orbits of E is finite (e.g. if E is defined by an algebraic Pfaffian form), then E has a first integral of class C^r.

If every germ of functions of class C^r on the space V of orbits of E can be represented by a C^r-function on V then there is a submersion of class C^r of V into $\mathbb{R}$. Hence there exists a C^r integral for E without a critical point.

In particular this last remark allows us to show that every C^r direction field has a C^r integral without critical point on every relatively compact open set (theorem of Kamke).

ii) The direction field defined by the algebraic Pfaffian form $\left[2(xy+1)+x^3y\right]dx + x^4dy$ has the function $(xy+1)\exp(-1/x^2)$ as an integral of class C^∞. But it has no analytic integral.

iii) Riemann's theorem on conformal mappings allows the extension of the results of this section to all smooth direction fields on simply connected open sets of the plane.

6. DIRECTION FIELDS ON CYLINDERS AND MOEBIUS STRIPS

The following lemma is essential for understanding this section.

6.1. LEMMA. Let J denote a Jordan curve on the cylinder $C = \mathbb{R}^2 - \{0\}$. If the interior of J contains the origin of $\mathbb{R}^2$ then J represents a generator of the fundamental group of C; otherwise J is null homotopic in C.

Indeed the index of the vector field $X: u \longmapsto u$ along J is ± 1 in the first case, and 0 in the second case (theorems 4.7 and 4.18).

We conclude:

6.2. PROPOSITION. A smooth direction field on the cylinder C having a periodic orbit or a closed transversal is orientable.

By I-5.2 such a periodic orbit or closed transversal represents indeed a generator of $\Pi_1(C)$.

6.3. COROLLARY. Every direction field tangent to the boundary (respectively transverse to it) on the annulus $S^1 \times [0,1]$ is orientable.

6.4. Remark. Since every two-sheeted covering space of the cylinder C is still diffeomorphic to C, the study of a direction field on C is easily reduced to the study of a vector field on C.

6.5. Exercises.

i) A smooth direction field on C without a periodic orbit has a closed orbit.

ii) In order to investigate a second order differential equation of the type $\theta'' + f(\theta)\,\theta' - g(\theta) = 0$ where f and g are periodic of period 2π one can associate to it a vector field X of components $P = z$ and $Q = -f(\alpha)z + g(\alpha)$ on the cylinder $S^1 \times \mathbb{R}$. If f and g are strictly positive the above equation has one and only one periodic solution: note that Q is positive for $z \leq 0$, and negative for large z, and a periodic orbit of X can be parametrised in the form $z = z(\alpha)$ with

$$\int_0^{2\pi} \left[\frac{g(\alpha)}{z(\alpha)} - f(\alpha)\right] d\alpha = 0 \,.$$

6.6. <u>The Moebius strip</u>. Let the involution $\alpha : (\theta, t) \longmapsto (\theta + \pi, -t)$ act on the annulus $\Gamma = S^1 \times [-1, +1]$, and denote the Moebius strip thus obtained as quotient space by B. The interior M of B is an open Moebius strip.

A Jordan curve J in M is nullhomotopic, or represents either a generator or the square of a generator of the fundamental group of M. Hence the complement of J with respect to M has two components (one of which is a disc) in the first case, and either one or two components in the second case. (They are either a cylinder, or a cylinder and an open Moebius strip.) The proof of the Schoenflies theorem (4.13) may be adapted to show that the quotient space of a closed annulus C by a smooth involution α without a fixpoint and reversing the orientation is diffeomorphic to a Moebius strip. (In the interior of C there is a Jordan curve which remains invariant under α.)

Every direction field on B has therefore at most one periodic orbit representing a generator of its fundamental group, and it is orientable if there is such an orbit.

It is also feasible to speak of components of type I, II, or III for a vector field without singularity on B (cf.2.8 and 2.10).

6.7. Examples.

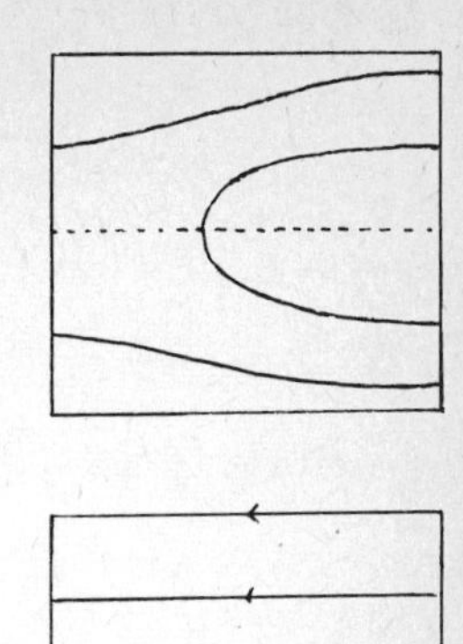

i) The vector field with components $P = 1$ and $Q = t(t^2 - 1)$ (respectively $P = t^2 - \frac{1}{4}$ and $Q = t(t^2 - 1)$) on Γ is invariant under the involution α. It determines a vector field $\hat{Y}_1$ (respectively $\hat{Y}_2$) on B which is tangential at the boundary, has no singularity, and has a single periodic orbit in M. This orbit represents a generator of $\Pi_1(B)$, and it is a stable limit cycle.

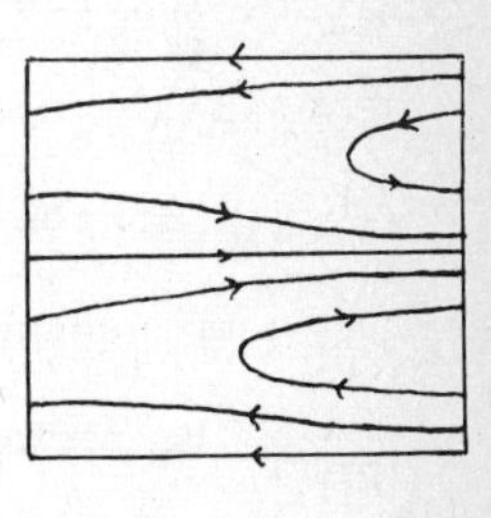

ii) The vector field with components $P = 1$ and $Q = 0$ on Γ is invariant under α. It determines a vector field $\hat{Z}$ on B, tangential at the boundary and without singular point, and with all its orbits periodic. (One of them represents a generator of $\Pi_1(B)$.)

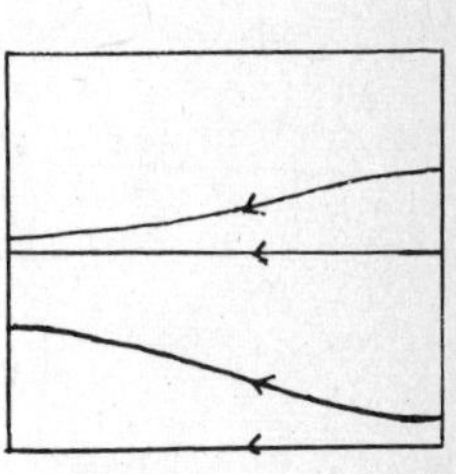

iii) The direction field corresponding to the vector field with components $P = t$ and $Q = t^2 - 1$ on Γ is invariant under α. It determines a non-orientable direction field F on B having the boundary of B as its only periodic orbit.

The following proposition shows the importance of these four examples.

6.8. PROPOSITION. Let E be a smooth direction field on the Moebius strip B which is tangential at the boundary.

Let E have a single periodic orbit in the interior of B representing a generator of $\Pi_1(B)$. Then E is orientable and topologically conjugate to the direction field of the vector field $\hat{Y}_1$ or $\hat{Y}_2$.

If all orbits of E are periodic, then E is orientable and differentiably conjugate to the direction field of the vector field Z.

If E has no periodic orbit in the interior of B then it is topologically conjugate to the direction field F.

In these cases B is called a <u>component of type I', II', III' or IV'</u> for E.

<u>Proof</u>. The inverse image of E under the projection of Γ onto B is a direction field which is tangential at the boundary and orientable.

In the first case it has two components of type I or II, a single component of type III in the second case, and a component of type II in the third case. The conjugations constructed in the proofs of theorem 2.8 and of proposition 2.10 may then be modified to become invariant under the involution α. Q.E.D.

We also have (cf. proposition 2.5):

6.9. <u>PROPOSITION</u>. Every orientable direction field on B which is transverse to the boundary has a periodic orbit.

On the other hand this result fails for a non-orientable direction field, as shown by the example of the direction field on B whose orbits are the projections of the generating lines of Γ.

6.10. <u>Application: Direction field on the Moebius strip</u> (cf. Kneser [6]).

Let E be a smooth direction field on the Moebius strip B

which is tangential at the boundary. The inverse image on Γ of E under the projection from Γ to B is an orientable direction field which is invariant under α.

Hence we have (cf. 2.11):

i) All orbits of E are proper, and have a periodic orbit as α-limit and ω-limit set.

ii) The set of periodic orbits of E is closed.

iii) The set of components of type II in B is finite.

iv) B contains at most one component of type I', II', III' or IV'.

v) If B does not contain a component of type IV', then E is orientable.

vi) If B does not contain any component of the types II, II', or IV', then E is differentiably conjugate to the direction field of the suspension vector field of a decreasing diffeomorphism of the segment $[-1,+1]$. In particular E has a periodic orbit representing a generator of $\Pi_1(B)$.

vii) If E is analytic then B is either a component of type III', or it is a finite union of components of types I, I', II, II' and IV'.

APPENDIX: SINGULAR GENERIC FOLIATIONS OF A DISC

In the theory of foliations one comes across vector fields on the disc $\mathbb{D}^2$ which are transverse to the boundary and whose singular points are saddles or centres (in the sense of section 3), and such that none of their orbits has two different saddles as limit sets.

Such fields define (oriented) "foliations with generic singularities" of $\mathbb{D}^2$. We are going to exhibit here the behaviour of such a field X which we assume to point inward on the boundary S^1 of $\mathbb{D}^2$.

A.1. PROPOSITION. The field X has one centre more than the number of its saddles.

By §4 the index of X along S^1 is indeed +1. The index of a centre is +1, and of a saddle -1.

A.2. PROPOSITION. The ω-limit set of a non-singular orbit γ of X is of one of the following four types:

i) a saddle;

ii) a periodic orbit;

iii) the union of a saddle and of a non-singular orbit having this saddle as α-limit and ω-limit sets;

iv) the union of a saddle and of two non-singular orbits having this saddle as α-limit and ω-limit sets.

Analoguous results hold for the α-limit sets of orbits which do not intersect S^1.

Proof. Since a centre is never in a limit set of a non-singular orbit we may assume that the limit set Ω_γ is neither a saddle nor a periodic orbit. This set is then compact, connected, and non-empty. Each of its non-singular orbits has as α-limit and ω-limit sets a common saddle, because of the third assumption about X. But then Ω_γ contains a single saddle and at most two non-singular orbits. Q.E.D.

The following sketches illustrate the situations iii) and iv).

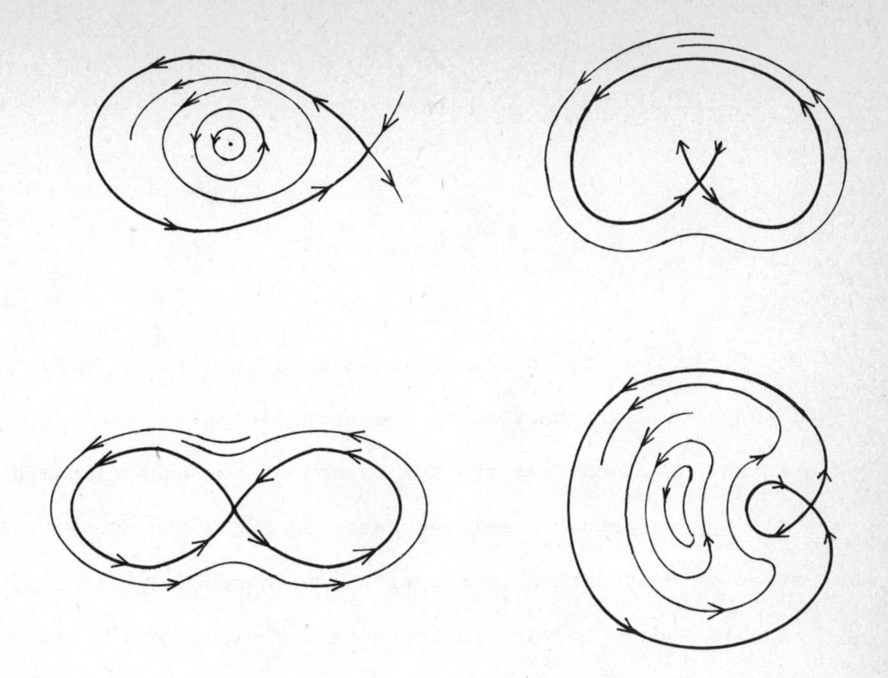

A.3. Remark. Except in case i) of A.2 the set of orbits of X having the same ω-limit set as γ is open in $\mathbb{D}^2$: If T is a transverse arc to X intersecting γ and with its initial point u at a regular point of Ω_γ, and if u_1, u_2, u_3 are three successive intersection points of γ with T (cf. exercise i) of 1.8) then every orbit intersecting T between u_1 and u_2 cuts it again between u_2 and u_3.

This procedure allows in particular to define on T a "Poincaré map" for Ω_γ. It will be a contraction.

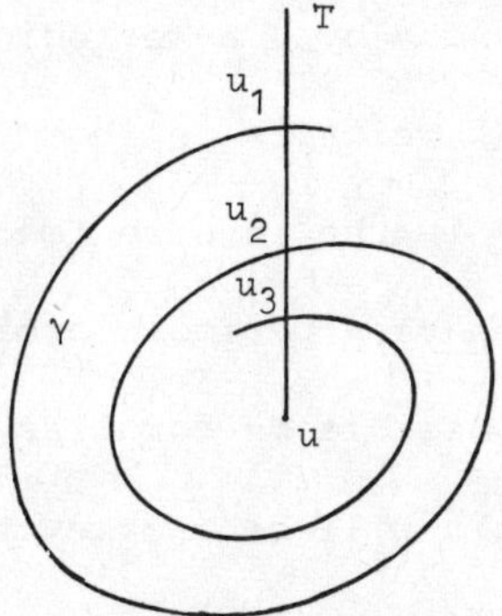

Chapter IV. Direction Fields on the Torus and Homeomorphisms of the Circle

The first results on this subject are again due to H. Poincaré who showed in particular the rôle of the homeomorphisms of the circle in this context. He discovered the phenomenon of the exceptional homeomorphisms which was further clarified later on by A. Denjoy, as well as the properties of conjugation to rotations which recently were continued by V. Arnold and M. Herman. Furthermore the global qualitative description of direction fields on the torus and on the Klein bottle are due to H. Kneser.

1. DIRECTION FIELDS ON THE TORUS [6]

Let E denote a smooth direction field on the torus $T^2 = \mathbb{R}^2/\mathbb{Z}^2$.

1.1. PROPOSITION. A periodic orbit of E (respectively a closed transversal to E) is a Jordan curve which is not nullhomotopic on T^2.

Let us indeed denote by J a periodic orbit of E (respectively a closed transversal to E) which is nullhomotopic on T^2. Then the direction field on $\mathbb{R}^2$ which is the inverse image of E under the projection of $\mathbb{R}^2$ onto T^2 would have a periodic orbit (respectively a closed transversal). This is impossible by corollary III-2.3.

This simple result will be the key to our investigation by

means of the following lemma:

1.2. LEMMA. Let J be a non nullhomotopic Jordan curve on T^2. Then the homotopy class j of J in the fundamental group $\Pi_1(T^2)$ is represented, under the canonical identification with $\mathbb{Z}\times\mathbb{Z}$, by an ordered pair (m,n) of relatively prime integers. The covering of T^2 corresponding to the subgroup of $\Pi_1(T^2)$ generated by j is isomorphic to the covering of T^2 by the cylinder $S^1\times\mathbb{R} = \mathbb{R}^2/\mathbb{Z}\times\{0\}$.

Proof. If m and n have a common factor $r\geq 1$ then the covering C of T^2 corresponding to the subgroup of $\Pi_1(T^2)$ generated by the element $(\frac{m}{r}, \frac{n}{r})$ is diffeomorphic to the cylinder, and an inverse image of J in C is a Jordan curve representing r times a generator of $\Pi_1(C)$. By lemma III-6.1 we conclude $r = 1$.

There is then a unimodular transformation H of $\mathbb{R}^2$ which induces a diffeomorphism h of T^2 such that h(J) represents the element (1,0) of $\Pi_1(T^2)$. Hence H induces an isomorphism of the covering $C \to T^2$ onto the covering $S^1\times\mathbb{R} \to T^2$. Q.E.D.

1.3. Remarks.

i) Let m and n be two relatively prime integers. Then the projection of the segment from (0,0) to (m,n) in $\mathbb{R}^2$ is a Jordan curve J in T^2 which represents the element (m,n) of $\Pi_1(T^2)$. J is called a torus knot of type (m,n).

ii) The complement of a nullhomotopic Jordan curve J on T^2 has two components one of which is diffeomorphic to an open disc.

On the other hand the complement of a Jordan curve J which is not nullhomotopic on T^2 is connected and diffeomorphic to a cylinder.

1.4. <u>Application: Direction Field on the Torus with a Periodic Orbit</u>.

By means of lemma 1.2 the study of a direction field E having a periodic orbit is reduced to that of an orientable direction field on an annulus (cf. proposition III-6.2). With an extension of the concept of component to the case of T^2 we thus find (cf. III-2.11):

i) All orbits of E are proper, and they have a periodic orbit as α-limit and ω-limit set.

ii) Two periodic orbits of E are homologuous in T^2 (i.e. they represent the same element of $\Pi_1(T^2)$, or more exactly they bound an annulus).

iii) The set of periodic orbits of E is closed.

iv) The set of components of type II in T^2 is finite, and E is orientable if and only if this number is even.

v) If T^2 does not contain any component of type II then the field E is differentiably conjugate to the field of directions of the suspension vector field of a diffeomorphism of the circle (preserving the orientation and of the same smoothness as E).

vi) If E is analytic then all its orbits are periodic, or T^2 is a finite union of components of type I and II.

vii) The field E is differentiably conjugate to a direction field whose periodic orbits are circles $S^1 \times \{\theta\}$.

These properties give a complete description of the direction fields on the torus which have a periodic orbit. The following result reduces the study of the direction fields without a periodic orbit to homeomorphisms of the circle which we treat in the following sections.

1.5. <u>THEOREM</u>. A smooth direction field E on the torus T^2 having no periodic orbit is orientable and differentiably conjugate to the direction field of the suspension vector field of a diffeomorphism of a circle (preserving the orientation and in the same differentiability class as E).

The proof of this theorem makes use of the following lemma:

1.6. LEMMA. If the direction field E has no periodic orbit, then it has closed transversals, and every one of them intersects every half-orbit of E infinitely often.

Proof. Let J be a closed transversal to E (cf. proposition I-5.12). Since J is not nullhomotopic in T^2 its lifts into the corresponding cylinder C (cf. lemma 1.2) are homologuous Jordan curves, which are not nullhomotopic and transverse to the orientable direction field F obtained as inverse image of E under the projection from C to T^2. If there were a half-orbit of E which does not intersect J then there would exist a half-orbit of F not intersecting any of its lifts. The corresponding orbit would have a periodic orbit as one of its limit sets, which is contrary to the assumption. Q.E.D.

Proof of theorem 1.5. Continuing with the above notations let J_1 be a lift of J in C, and J_2 the lift obtained from J_1 by one of the generators τ of the automorphism group of C. Let also h(u) denote the intersection with J_2 of the orbit of F passing through the point u of J_1. Then h is a diffeomorphism of J_1 onto J_2 (with the same smoothness as E), and E is differentiably conjugate to the direction field of the suspension field of the diffeomorphism $\tau^{-1} \circ h$ of J_1. Q.E.D.

1.7. Milnor's invariant of direction fields on T^2.

Let E be a smooth direction field on T^2, and denote by F its inverse image on $\mathbb{R}^2$ under the projection q of $\mathbb{R}^2$ onto T^2. Furthermore let K be a compact subset of $\mathbb{R}^2$ whose interior covers T^2 (i.e. $q(\mathring{K})=T^2$). We identify $\Pi_1(T^2)$ with the subgroup $\mathbb{Z}^2$ of $\mathbb{R}^2$ and denote then by π_K the set of non-vanishing elements τ of $\Pi_1(T^2)$ for which there exists an orbit of F intersecting K and its translate $\tau(K)$. Furthermore we denote by R_K the subspace of the projective line $\mathbb{P}\mathbb{R}^1$ corresponding to the cone over π_K in $\mathbb{R}^2$, and by ρ_K the compact set of cluster points of R_K in $\mathbb{P}\mathbb{R}^1$.

For a second compact subset K' of $\mathbb{R}^2$ whose interior covers T^2 there is a finite family $\tau_1,\ldots,\tau_n$ of elements of $\mathbb{Z}^2$ such that the union $\bigcup_{i=1}^{n}\tau_i(K')$ contains K. For every $\tau \in \pi_K$ there are two of these translations τ_α and τ_β such that $\tau_\alpha^{-1}\tau\,\tau_\beta \in \pi_{K'}$. If (τ_i) is then a sequence of distinct elements of π_K which determine a limiting direction Δ of ρ_K then the corresponding sequence $(\tau_{\alpha_i}^{-1}\tau_i\,\tau_{\beta_i})$ of $\pi_{K'}$ determines the same limit Δ. Hence the compact set ρ_K is independent of the choice of the compact set K. It will be denoted by $\rho(E)$.

1.8. Example. Let E be the direction field on T^2 corresponding to the field F on R^2 whose orbits are the lines parallel to a direction Δ. Then the invariant $\rho(E)$ consists of the point of PR^1 corresponding to Δ.

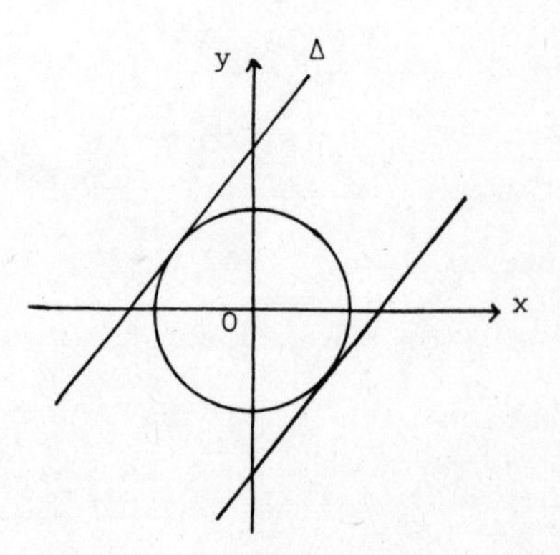

1.9. PROPOSITION. Let h denote a topological conjugation of E to a smooth direction field E', and $\tilde{h}$ the diffeomorphism of $\mathbb{P}\mathbb{R}^1$ induced by the unimodular transformation of $\mathbb{R}^2$ which is determined by the automorphism h_* of $\Pi_1(T^2)$. Then $\rho(E') = \tilde{h}(\rho(E))$.

Let indeed $H: \mathbb{R}^2 \rightarrow \mathbb{R}^2$ be the lift of h and $K' = H(K)$. Then $\tau_{K'} = h_*(\tau_K)$, because an orbit γ of E intersecting K and $\tau(K)$ produces an orbit $H(\gamma)$ of F' intersecting K' and $(h_*\tau)(K') = H(\tau(K))$.

1.10. COROLLARY. Let h be a topological conjugation from E to E' which is homotopic to the identity. Then $\rho(E) = \rho(E')$.

1.11. Examples.

i) Let E be a direction field on T^2 having a periodic orbit. Then its invariant $\rho(E)$ is a "rational" point of $\mathbb{P}\mathbb{R}^1$, i.e. it corresponds to a direction with a rational slope (property vii) of 1.4).

ii) Let E be topologically conjugate to the direction field of the suspension field of the rotation through the angle α of the circle $S^1 = \mathbb{R}/\mathbb{Z}$. Then the invariant is a point on the orbit of the direction of slope α under the group of integral projective transformations, i.e. its slope is $(p+q\alpha)/(m+n\alpha)$ where m,n,p,q are integers satisfying $|mq-np| = 1$.

From corollary 5.4 we can thus deduce:

1.12. THEOREM. Let E be a direction field of class C^2 on T^2. Then its invariant $\rho(E)$ is a point of $\mathbb{P}\mathbb{R}^1$ which is rational or irrational according to whether E has or has not a periodic orbit.

1.13. THEOREM. Let E and E' be two direction fields of class C^2 and having no periodic orbits on T^2. They are topologically conjugate if and only if their invariants $\rho(E)$ and $\rho(E')$ belong to the same orbit of the group of projective transformations with integral coefficients.

1.14. Exercise. Let ω be a closed Pfaffian form without singularity on T^2, and let E be the corresponding direction field.

i) If X is a smooth vector field on T^2 satisfying $\omega(X) = 1$ then the flow generated by X leaves ω invariant. Hence the direction field either has only periodic orbits or none.

ii) Multiplying ω by a constant, if necessary, we can find a closed transversal J to E intersecting all orbits of E and having a parametric representation $c: S^1 \longrightarrow T^2$ satisfying $c^*\omega = d\theta$.

iii) Let $P: J \to J$ be the Poincaré map corresponding to the transversal J. If the group of the periods of ω is generated by the numbers 1 and α, then the map $c^{-1} \circ P \circ c$ is the rotation of S^1 through the angle α. Hence E is differentiably conjugate to the constant direction field of slope α.

iv) If the group of periods of ω is of rank 1, then all orbits of E are periodic; if it is of rank 2 they are all everywhere dense.

1.15. Exercise. An orientable direction field E on a subspace A of T^2 corresponds to a map of A into S^1. In particular for $A = \gamma$ = a Jordan curve on T^2 we may define the index $i_E(\gamma)$ of E along γ as in the plane.

i) If in addition γ is tangent (or transverse) to E the index $i_E(\gamma)$ equals +1 or 0 according to whether γ is or is not nullhomotopic in T^2.

ii) Let E be a smooth direction field having no periodic orbit on T^2 and corresponding to a map $h: T^2 \to S^1$. Then the homomorphism $h_*: \Pi_1(T^2) \to \Pi_1(S^1)$ is trivial.

2. DIRECTION FIELDS ON A KLEIN BOTTLE (cf. [6]).

2.1. The Klein bottle. Let G be the group of transformations of the plane $\mathbb{R}^2$ generated by the translation $h: (u,v) \mapsto (u+1,v)$ and by the map $k: (u,v) \mapsto (-u,v+1)$. This group acts properly and freely on the plane, and the quotient space of $\mathbb{R}^2$ by this action is the Klein bottle K^2. It is a compact and non-orientable surface with fundamental group isomorphic to G. K^2 may be visualized as the space obtained from a square by an identification of opposite sides as shown in the sketch below.

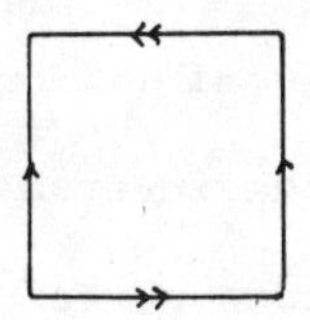

Because of $hk = kh^{-1}$ every element of G may be uniquely written as $k^m.h^n$, with $m,n \in \mathbb{Z}$. The rules of computation in G are then

$$(k^m h^n)(k^p h^q) = k^{m+p} h^{(-1)^p n+q},$$
$$(k^m h^n)^{-1} = k^{-m} h^{(-1)^{m+1} n}.$$

In particular the conjugacy class of an element $k^m h^n$ of G for m even is $\{k^m h^n, k^m h^{-n}\}$, and $\{k^m h^{n+2p} | p \in \mathbb{Z}\}$ for m odd. The subgroup H of G generated by h and k^2 is normal and of index 2. The covering T of K^2 corresponding to this subgroup is thus a normal two-sheeted covering space which is diffeomorphic to the torus T^2.

2.2. PROPOSITION. Let E denote a smooth direction field on the Klein bottle K^2. A periodic orbit of E (respectively a closed transversal to E) is a non nullhomotopic Jordan curve on K^2.

The proof of this result is identical to the one of proposition 1.1 .

2.3. LEMMA. A non nullhomotopic Jordan curve on K^2 represents one of the following elements of the fundamental group:

$$h,\ h^{-1},\ k^2,\ k^{-2} \quad \text{or } k^{-1}h^n \quad \text{with } n \in \mathbb{Z}\ .$$

In view of the determination of the conjugacy classes in G this result is independent of the choice of the base point for $\Pi_1(K^2)$.

Proof. Let j be the homotopy class if J in $\Pi_1(K^2)$. If $j = k^{2m+1}h^n$ is not in H, the inverse image of J in T is a Jordan curve representing the element $(k^{2m+1}h^n)^2 = k^{4m+2}$ of $\Pi_1(T)$. By lemma 1.2 one finds then $m=0$ or $m=-1$.

If $j = k^{2m}h^n$ is in H, then the inverse image of J in T consists of two homologuous and disjoint Jordan curves J_1 and J_2 which are interchanged by the involution α of the covering. The integers m and n are thus relatively prime by lemma 1.2, and the complement of $J_1 \cup J_2$ in T has two components U_1 and U_2.

If α leaves U_1 invariant the projection into K^2 of the closure $\bar{U}_1 = U_1 \cup J_1 \cup J_2$ of U_1 is a Moebius strip having J as boundary. j is then a square in G, from which follows $j = k^2$ or $j = k^{-2}$.

If α interchanges U_1 and U_2 we can find a Jordan curve I in T which is invariant under α and represents an element $i = k^{2p}h^q$ of $\Pi_1(T)$, with $mq-np = 1$. But then i is a square in G, and $q = 0$ and $p = -n = \varepsilon$, $\varepsilon = \pm 1$. The elements kj and $k^{-1}j$ can also be represented by a Jordan curve in K^2, and hence $m = 0$. Q.E.D.

2.4. Remarks.

i) The sketches below show how these homotopy classes can be realized by Jordan curves.

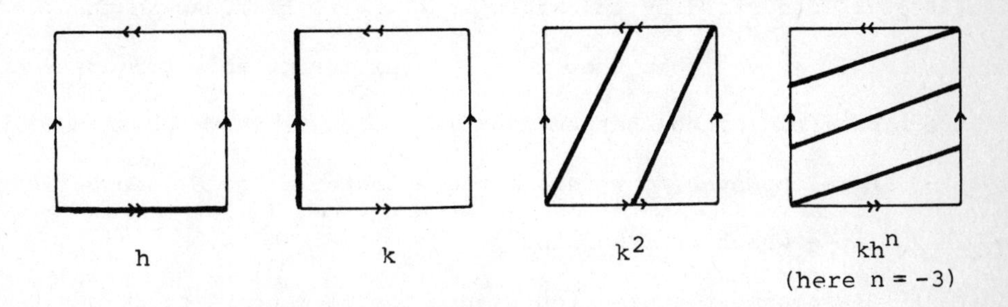

ii) The Jordan curve J is two-sided if its homotopy class is $h^{\pm 1}$ or $k^{\pm 2}$, otherwise one-sided (cf. exercise I-5.13). Its tubular neighbourhood is thus a cylinder in the first cases, and a Moebius strip in the other cases.

iii) Let J be a Jordan curve representing the element j of $\Pi_1(K^2)$. Then its complement has

- 2 components for $j = 1$ (one of which is diffeomorphic to an open disc);
- 1 component diffeomorphic to a cylinder for $j = h^{\pm 1}$;
- 1 component diffeomorphic to a Moebius strip for $j = kh^n$ or $j = k^{-1}h^n$;
- 2 components both diffeomorphic to a Moebius strip for $j = k^{\pm 2}$.

2.5. THEOREM (Kneser). Every smooth direction field on the Klein bottle K^2 has a periodic orbit.

Proof. Let E denote a smooth direction field on K^2 without a periodic orbit, and J a closed transversal to E (cf. proposition I-5.12).

Assume first of all that the homotopy class of J in $\Pi_1(K^2)$ is h. Then the covering of K^2 corresponding to the subgroup generated by h is normal and diffeomorphic to the cylinder $S^1 \times \mathbb{R}$, with an automorphism group generated by the map $k:(\theta,t) \mapsto (-\theta,t+1)$. As in the proof of theorem 1.5 one can then show that E is differentiably conjugate to the direction field of the suspension field of a diffeomorphism φ of the circle S^1. Here, however, φ reverses the orientation of S^1. Hence it has a fixpoint - which is impossible.

Assume next that J represents the element k^2 of $\Pi_1(K^2)$. Again the covering Γ of K^2 correponding to the subgroup generated by k^2 is normal and diffeomorphic to the cylinder $\mathbb{R} \times (\mathbb{R}/2\mathbb{Z})$, with an automorphism group generated by the translation $h:(t,\theta) \mapsto (t+1,\theta)$ and the map $\varkappa:(t,\theta) \mapsto (-t,\theta+1)$. The lifts of J in Γ are mutually disjoint, homologuous and non nullhomotopic Jordan curves which are transverse to the inverse image direction field F of E under the projection of Γ onto K^2. They form two families which are invariant under h and are interchanged by $\varkappa$.

Then we can find three elements J_o, $J_1 = \varkappa(J_o)$, and $J_2 = h(J_o)$ such that J_1 lies between J_o and J_2. The annulus bounded in Γ by J_o and J_1 (respectively J_1 and J_2) is invariant under $\varkappa$ (respectively $\varkappa h^{-1}$). Since F has no periodic orbit the diffeomorphism φ (respectively Ψ) of J_o onto J_1 (repsectively J_1 onto J_2) obtained by following the integral curves of F satisfies $\varphi\varkappa\varphi = \varkappa$ (respectively $\Psi\varkappa h^{-1}\Psi = \varkappa h^{-1}$). The inverse image $\hat{F}$ of F on the torus $T = \Gamma/(h)$ under the projection of Γ onto T is differentiably conjugate to the direction field of the suspension field of the diffeomorphism $\bar{\omega} = h^{-1} \circ \Psi \circ \varphi$ of J_o.

This diffeomorphism $\bar{\omega}$ is, however, the product of the diffeomorphisms $\varkappa \circ \varphi$ and $h^{-1} \circ \Psi \circ \varkappa$ of J_o. Both of them preserve the orientation and their squares equal the identity map. By proposition 5.7 and remark ii) of 4.7 we recognize that either $\bar{\omega}$ or $\bar{\omega}^2$ must have a fixpoint. Again this is contradictory.

Finally, if the homotopy class of J in $\Pi_1(K^2)$ is kh^n, $n \in \mathbb{Z}$, there is another closed transversal to E representing the element $(kh^n)^2 = k^2$. This reduces everything to the preceding case. Q.E.D.

2.6. <u>COROLLARY</u>. Every orbit of a smooth direction field on a Klein bottle K^2 is proper.

By 1.4 the corresponding direction field on the torus T^2 has indeed a periodic orbit.

Hence we deduce:

2.7. <u>PROPOSITION</u>. A minimal set of a smooth direction field on a Klein bottle is a periodic orbit.

2.8. <u>Application: Direction Field on a Klein Bottle</u>. Let E be a smooth direction field on the Klein bottle K^2, and denote by J a periodic orbit of E (theorem 2.5). We have to consider three cases according to whether J represents the element h or h^{-1}, k^2 or k^{-2}, kh^n or $k^{-1}h^n$, $n \in \mathbb{Z}$ of $\Pi_1(K^2)$.

If J represents the element h of $\Pi_1(K^2)$, the investigation may be reduced to considering an orientable direction field on an annulus, by passing to the normal covering space corresponding to the subgroup generated by h. We obtain (cf. III-2.11):

i) All orbits of E are proper, and they have a periodic

orbit as α-limit and ω-limit set.

ii) Two periodic orbits of E are homologuous in K^2 and bound together an annulus.

iii) The set of periodic orbits of E is closed.

iv) K^2 is the union of components of types I, II, and III.

v) The set of components of type II is finite. E is orientable if and only if their number is odd.

vi) If E is analytic either all its orbits are periodic, or K^2 is a finite union of components of types I and II.

Let now J represent the element k^2 of $\Pi_1(K^2)$. Again passing to the normal covering space corresponding to the subgroup generated by k^2 we may study E by considering a direction field on the union of two Moebius strips each having J as its boundary (cf. proof of theorem 2.5). We may thus list the following properties (cf. III-6.10):

i) All orbits of E are proper and have a periodic orbit as their α-limit and ω-limit set.

ii) Every periodic orbit, except possibly one or two, is homologuous to J in K^2 and bounds together with J an annulus. Every exceptional periodic orbit represents an element of $\Pi_1(K^2)$ whose square is k^2 or k^{-2}.

iii) The set of periodic orbits of E is closed.

iv) K^2 is the union of components of types I, II, III, I', II', III' and IV', with a finite number of components of type II and two components of types I', II', III', or IV'.

v) E is orientable if and only if K^2 contains no component of type IV'.

vi) If K^2 contains no component of types II, II' or IV',

then E is differentiably conjugate to the direction field of the suspension field of a diffeomorphism of S^1 reversing the orientation and of the same smoothness as E.

vii) If E is analytic all its orbits are either periodic, (i. e. K^2 is the union of two components of type III') or K^2 is a finite union of components of types I, II, I',II', and IV'.

Finally let J represent an element kh^n of $\Pi_1(K^2)$; and let again Γ denote the normal covering space corresponding to the subgroup generated by k^2. If J_o and $J_1 = hJ_o$ are two inverse images of J in Γ the annulus bounded by them is invariant under the automorphism kh^{n-1}; hence we may study a direction field on a Moebius strip. We may restrict ourselves to the case where E has at most two periodic orbits, each of which represents an element of $\Pi_1(K^2)$ whose square is k^2 or k^{-2} (in the other cases E has a periodic orbit whose homotopy class is k^2 or k^{-2}). We have then the following properties:

i) All orbits of E are proper, and they have a periodic orbit as α-limit and ω-limit set.

ii) K^2 consists of a single "component", which is of type I' or II' if E has two periodic orbits, and of type IV' if only one.

iii) E is orientable if and only if K^2 is a component of type I' or II'.

iv) If K^2 is a component of type I' then E is differentiably conjugate to the suspension field of a diffeomorphism of S^1 (which reverses the orientation, has the same smoothness properties as E, and whose square has only two fixpoints).

3. HOMEOMORPHISMS OF THE CIRCLE WITHOUT PERIODIC POINT.

Let f denote a homeomorphism of the circle $S^1 = \mathbb{R}/\mathbb{Z}$ without a periodic point, hence preserving the orientation, and let $p: \mathbb{R} \to S^1$ be the projection.

3.1. PROPOSITION. All orbits of f have a compact subset of S^1 as common α-limit and ω-limit set.

Proof. Let γ be an orbit of f, and assume first that its limit set Ω_γ (which is compact and non-empty) is different from S^1. If γ' is an orbit in Ω_γ its α-limit and ω-limit sets are contained in Ω_γ. If γ' is an orbit in $S^1 - \Omega_\gamma$ it has at most one point in each of the open arcs which make up $S^1 - \Omega_\gamma$: otherwise one of these components would be invariant under a power of f, and f would have a periodic point. Hence once more the limit sets of γ' are contained in Ω_γ, and all α-limit and ω-limit sets coincide with Ω_γ.

The same holds, of course, for $\Omega_\gamma = S^1$. Q.E.D.

3.2. Remark. A homeomorphism of S^1 preserving the orientation and having a periodic point of minimal period n is such that all of its periodic points have period n. A homeomorphism of S^1 reversing the orientation has exactly two fixpoints, and all other periodic points (if there are any) have period 2.

In both of these two cases every orbit has a periodic orbit as α-limit and ω-limit set.

3.3. COROLLARY. A homeomorphism f of S^1 without periodic points has a single minimal set.

This minimal set M coincides indeed with the common limit set of all orbits of f.

We are thus lead to distinguish the following two possibilities:

i) If the interior of M is non-empty, M is S^1, and the orbits of f are all everywhere dense. j is then called an <u>ergodic</u> homeomorphism of S^1 (cf. exercise 5.10).

ii) If the interior of M is empty M is perfect and totally disconnected, hence homeomorphic to a Cantor set. The orbits of f in $S^1 - M$ are proper, and those in M are exceptional. f is then called an <u>exceptional</u> homeomorphism of S^1, and M an <u>exceptional minimal set</u>.

If f is of class C^2 we will show that only the first situation can occur (theorem 3.4). The second possibility can present itself in the classes C^0 and C^1 (proposition 3.8 and exercise 3.10).

3.4. <u>THEOREM</u> (Denjoy [2]). A diffeomorphism of the circle S^1 of class C^2 and without periodic point is ergodic.

More precisely we will show that this result holds for f of class C^1 with a derivative f' of bounded variation.

<u>Proof</u> (following C.L.Siegel: Ann.of Math.46, 1945; see also the proof of theorem V-2.4).

Let u and v be two different points of S^1. We denote by uv the open oriented arc with initial point u and terminal point v on S^1, and for $w \in S^1$ and $k \in \mathbb{Z}$ we set $w_k = f^k(w)$.

Assume now that the diffeomorphism f is of class C^1 and has no periodic point, and denote by M its exceptional minimal set. Choose an arc uv among those which make up the complement of M in S^1. The arcs $u_k v_k$, $k \in Z$, are then mutually disjoint, and the series $\sum_{-\infty}^{+\infty} l_k$ is convergent, where l_k = length of $u_k v_k$.

Taking into account the relations

$$\frac{df^m}{dw}(w_o) = \prod_{k=1}^{m} f'(w_{m-k}), \quad \frac{df^{-m}}{dw}(w_o) = \prod_{k=1}^{m} \frac{1}{f'(w_{-k})}, \quad m > 0,$$

we can find, for every integer n, two points a and b on the arc uv such that

$$\frac{l_n}{l_o} = \prod_{k=1}^{n} f'(a_{n-k}) \quad \text{and} \quad \frac{l_{-n}}{l_o} = \prod_{k=1}^{n} \frac{1}{f'(b_{-k})} .$$

We conclude

$$\operatorname{Log} \frac{l_o^2}{l_n l_{-n}} < \sum_{k=1}^{n} \left| \operatorname{Log} f'(b_{-k}) - \operatorname{Log} f'(a_{n-k}) \right| .$$

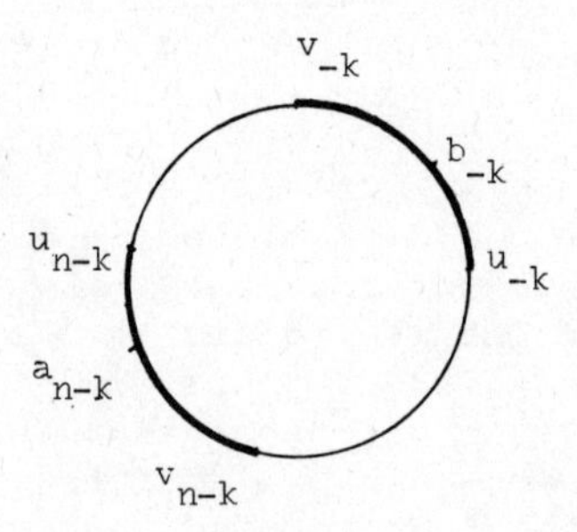

The left hand side of this inequality tends to infinity with n. If f' is of bounded variation on S^1, however, the same holds for Log f'. From the lemma below we deduce that for arbitrarily large values of n the right hand side is less than the variation of Log f' on

S^1. This will conclude the proof by contradiction. Q.E.D.

3.5. LEMMA. For every integer $N > 0$ there exists an integer $n > N$ such that one of the two families

$$\{u_{-n}v_o, u_{-n+1}v_1, \ldots, u_{-1}v_{n-1}\}, \quad \{u_o v_{-n}, u_1 v_{-n+1}, \ldots, u_{n-1}v_{-1}\}$$

consists of mutually disjoint arcs.

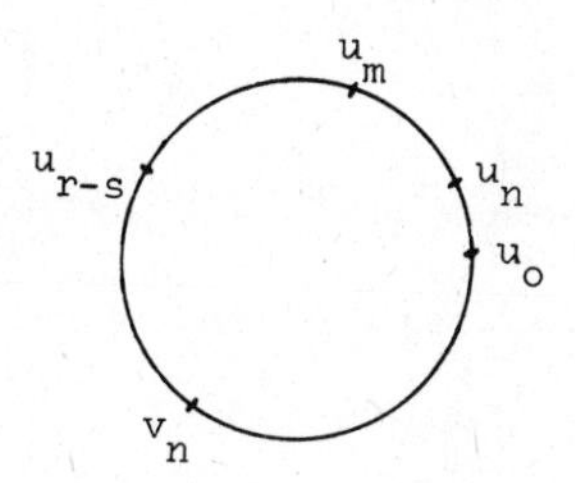

Proof. Let $u_o u_m$ denote the smallest among the arcs $u_o u_j$ for $0 < |j| \leq N$, and let n be the smallest among the integers $h > N$ such that one of the points u_h or u_{-h} belongs to the arc $u_o u_m$ (there are such h because otherwise the sequence of the points u_{mk}, $k \geq 1$ would be monotonic on the arc $u_m u_o$ and f would have a periodic point).

Assume now that for this integer n the two families mentioned in the lemma do not consist of mutually disjoint arcs. If $u_n \in u_o u_m$, there are two distinct integers r and s between 1 and n such that $u_{-s} \in u_{-r}v_{n-r}$. Hence the point u_{r-s} belongs to the arc $u_o v_n$, and the arcs $u_m v_n$ and $u_n v_n$ are not disjoint. If on the other hand $u_{-n} \in u_o u_m$ there are again two distinct integers r' and s' between 0 and n-1 with $u_{s'} \in u_{r'}v_{-n+r'}$. In both cases we arrive at a contradiction. Q.E.D.

3.6. COROLLARY. Every direction field of class C^r, $r \geq 2$, without periodic orbit on the torus T^2 is orientable, and all its orbits are everywhere dense.

3.7. COROLLARY. A minimal set of a direction field of class C^r, $r \geqslant 2$, on the torus T^2 is either a periodic orbit or coincides with T^2.

3.8. PROPOSITION. Let M be a closed, perfect, and totally disconnected subspace of S^1. Then there exists a homeomorphism of S^1 without periodic point having M as minimal set.

(This result was explicitly mentioned for the first time in A. Denjoy [2], but it was essentially known already by H. Kneser [6] and P.Bohl: Acta Math., 40, 1916).

Proof. Choose an irrational number α, and let r denote the rotation of S^1 corresponding to the translation $t \mapsto t+\alpha$ of $\mathbb{R}$. Every orbit of r is dense in S^1. For an orbit γ of r there is a bijection λ of γ onto the set Γ of components of the open set $U = S^1 - M$ which is compatible with "the order on S^1": i.e. if v is a point of γ between two other points u and w of γ, then the arc $\lambda(v)$ lies between the arcs $\lambda(u)$ and $\lambda(w)$.

Let then f be the diffeomorphism of U onto itself whose restriction to the arc $\lambda(u)$, $u \in \gamma$, of Γ is the affine and increasing bijection of this arc onto the arc $\lambda(r(u))$. This map is uniformly continuous: Given $\varepsilon > 0$ choose arcs $I_1, \ldots, I_q$ in Γ such that the component arcs of $S^1 - (I_1 \cup \ldots \cup I_q)$ all have length less than $\frac{1}{3}\varepsilon$. Then let a_k be the similarity ratio of f on the arc $J_k = f^{-1}(I_k)$, and choose $\delta > 0$ less than the numbers $\varepsilon/3a_k$ and less than the lengths of the components of $S^1 - (I_1 \cup \ldots \cup I_q)$.

The map f may be extended as a continuous map of S^1 into

itself which is a homeomorphism with the desired properties; in particular the subspace M is the ω-limit set of each orbit of f in U.

Q.E.D.

3.9. Remark. In the above examples the homeomorphism f acts transitively on the set Γ of the components of the open set $U = S^1 - M$. Replacing in the preceding construction the orbit γ of r by a countable (finite or infinite) union of orbits of r one obtains a homeomorphism f for which the open set U is a disjoint union (finite or infinite) of non-empty invariant open sets.

3.10. Exercise. We will present the construction of a diffeomorphism of the circle which is exceptional and of class C^1 (example of A. Denjoy [2], in a version of M. Peixoto).

For this, choose an irrational number α between 0 and 1, and a convergent series $\sum_{-\infty}^{+\infty} l_n$ with sum 1, whose terms are positive and satisfy

$$\lim_{n \to \infty} \frac{l_n}{l_{n+1}} = 1 \qquad \text{(cf. iv) of exercise 5.11).}$$

For every integer n denote

- by α_n the fractional part of $n\alpha$ (i.e. the number in the interval $[0,1)$ such that $n\alpha - \alpha_n$ is an integer),

- by A_n (respectively B_n) the set of integers q such that $\alpha_q < \alpha_n$ (respectively $\alpha_q \leq \alpha_n$),

- by a_n (respectively b_n) the number $\sum_{q \in A_n} l_q$ (resp. $\sum_{q \in B_n} l_q$),

- by J_n the interval $[a_n, b_n]$ of length l_n.

Moreover we choose for every integer n a continuous and strictly positive function φ_n on J_n with the following properties:

a) $\varphi_n(a_n) = \varphi_n(b_n) = 1$;

b) $\int_{a_n}^{b_n} \varphi_n(t)\,dt = l_{n+1}$;

c) $\lim_{n \to \infty} (\sup_{J_n} |\varphi_n - 1|) = 0$

One may, e.g., choose $l_n = \frac{\kappa}{1+n^2}$ and $\varphi_n(t) = 1 + k_n(t-a_n)(b_n-t)$.

We then have

i) The α_n are mutually distinct and dense on the interval $I = [0,1]$.

ii) If m and n are two integers such that $\alpha_m < \alpha_n$, then $a_m < b_m < a_n < b_n$.

Hence the intervals J_n are mutually disjoint, and the union of their interiors is an open and dense set in I.

iii) There exists a continuous, strictly positive, and periodic function φ of period 1 on $\mathbb{R}$, whose restriction to each interval J_n equals φ_n, and $\int_0^1 \varphi(t)\,dt = 1$.

iv) The indefinite integral F of φ satisfying $F(0) = a_1$ is an increasing C^1-diffeomorphism of $\mathbb{R}$ for which $F(t+1) = F(t)+1$ holds.

v) If $\alpha + \alpha_n < 1$ we have $F(J_n) = J_{n+1}$; if $\alpha + \alpha_n > 1$ then $F(J_n) = 1 + J_{n+1}$. (The set of integers q such that $q-1 \in A_n$ is $A_{n+1} - A_1$ in the first case, and $(\mathbb{Z} - A_1) \cup A_{n+1}$ in the second.)

vi) The diffeomorphism F induces a C^1-diffeomorphism f of the circle S^1 which preserves the orientation and satisfies the following properties:

The projection of the union of the intervals (a_n, b_n) onto S^1 is an open and dense subset of S^1 which is invariant under f. The complement M of U in S^1 is a closed, perfect, and totally disconnected set which is invariant under f. Every orbit of f in U has M as limit sets, hence f is an exceptional diffeomorphism.

4. ROTATION NUMBER OF POINCARE

4.1. The homeomorphism group of the circle. Let f be a homeomorphism of the circle S^1 preserving the orientation. A lift $F: \mathbb{R} \to \mathbb{R}$ of f is a map such that $f \circ p = p \circ F$; it is a strictly increasing function satisfying $F(t+1) = F(t) + 1$.

Beside the group $\mathcal{H}$ of homeomorphisms of S^1 preserving orientation we are thus lead to consider also the space $\mathcal{P}$ of continuous functions of period 1 on $\mathbb{R}$, the group $\mathcal{K}$ of increasing homeomorphisms of $\mathbb{R}$ of the form $Id + \Phi$, with $\Phi \in \mathcal{P}$, and the homeomorphism π of $\mathcal{K}$ onto $\mathcal{H}$ associating to an element F of $\mathcal{K}$ the homeomorphism of S^1 having F as its lift.

The kernel of π is the group of translations with integral shifts. This is also the center of $\mathcal{K}$.

The compact-open topology on $\mathcal{P}$ induces a topology on $\mathcal{K}$ such that $\mathcal{K}$ becomes a topological group. It corresponds to the metric

$$d(F,G) = \|F-G\| = \sup_{t \in \mathbb{R}} |F(t)-G(t)|$$

The center C of $\mathcal{K}$ is then a discrete subgroup of $\mathcal{K}$, and we identify the group $\mathcal{H}$ with the topological quotient group $\mathcal{K}/C$.

4.2. PROPOSITION. For every homeomorphism $F \in \mathcal{K}$ the sequence of periodic functions $\frac{1}{n}(F^n - Id)$, $n > 0$, converges in $\mathcal{P}$ (i.e. uniformly on $\mathbb{R}$) towards a constant $\rho(F)$.

The proof of this proposition makes use of the following lemma:

4.3. LEMMA. For an element $F = Id + \Phi$ of K we have $\sup_{\mathbb{R}} \Phi - \inf_{\mathbb{R}} \Phi < 1$.

Indeed if we had, by contradiction,

$$F(y) - y = F(x) - x + 1, \quad x < y < x+1, \quad \text{then} \quad F(y) > F(x) + 1 = F(x+1).$$

Proof of proposition 4.2. For every integer $q > 0$ we set

$$\alpha_q = \inf_{\mathbb{R}} (F^q - Id), \quad \beta_q = \sup_{\mathbb{R}} (F^q - Id).$$

Starting from the inequalities

$$\alpha_q \leqslant F^{qs}(t) - F^{q(s-1)}(t) \leqslant \beta_q, \quad s \geqslant 1,$$

we obtain, for an integer $n = mq+r$, $m \geqslant 0$, $0 \leqslant r < q$, the new inequalities

$$m\alpha_q + r\alpha_1 \leqslant F^n(t) - t \leqslant m\beta_q + r\beta_1,$$

$$\frac{\alpha_q}{q} \leqslant \liminf_{n \to +\infty} \frac{1}{n} \left[F^n(t) - t\right] \leqslant \limsup_{n \to +\infty} \frac{1}{n} \left[F^n(t) - t\right] \leqslant \frac{\beta_q}{q}.$$

Finally we apply lemma 4.3. Q.E.D.

4.4. COROLLARY. For every $t \in \mathbb{R}$, we have $\lim_{n \to +\infty} \frac{1}{n} F^n(t) = \rho(F)$.

4.5. Examples.

i) For a translation $R_\alpha : t \mapsto t+\alpha$ of $\mathbb{R}$, we have $\rho(R_\alpha) = \alpha$.

ii) For example 3.10 we obtain $\rho(F) = \alpha$: letting $[n\alpha]$ denote the integral part $n\alpha - \alpha_n$ of α_n, we have indeed $F^{n+1}(0) = [n\alpha] + a_{n+1}$ if $\alpha_n + \alpha < 1$, and $F^{n+1}(0) = [n\alpha] + 1 + a_{n+1}$ otherwise. Then note that

$$\lim_{n \to +\infty} \frac{1}{n} [n\alpha] = \alpha.$$

4.6. Exercises.

i) Let F denote a homeomorphism in $\mathcal{K}$. The subsets

$$R_- = \left\{\frac{n}{m} \in \mathbb{Q} \mid m > 0 \text{ and } F^m(0) > n\right\},$$
$$R_+ = \left\{\frac{n}{m} \in \mathbb{Q} \mid m > 0 \text{ and } F^m(0) < n\right\}$$

determine a Dedekind cut in $\mathbb{Q}$ corresponding to the number $\rho(F)$.

ii) Let Y denote a smooth vector field on $\mathbb{R}^2$ with components 1 and Q, with $Q(u,v)$ a function of period 1 in each variable. By

$$H = (t,u,v) \longmapsto (u+t, h(u,v,t))$$

we denote the one-parameter group generated by Y. Y is then complete. The map $F: t \mapsto h(0,t,1)$ is an element of $\mathcal{K}$, and for every point (u,v) we have

$$\rho(F) = \lim_{n \to +\infty} \frac{1}{n} F^n(t) = \lim_{t \to +\infty} \frac{h(u,v,t)}{u+t}.$$

In other words: the number $\rho(F)$ appears here as the "limit slope" of all integral curves of Y.

Note also that Y determines a vector field on the torus T^2 which is differentiably conjugate to the suspension field of the diffeomorphism $\pi(F)$ of S^1. Thus $\rho(F)$ may be compared to the invariant introduced in 1.7.

4.7. Remarks.

i) For $n=-m$, $m > 0$, we have

$$\frac{1}{n}\left[F^n - \mathrm{Id}\right] = \frac{1}{m}\left[F^m - \mathrm{Id}\right] \circ F^n,$$

hence

$$\rho(F) = \lim_{n \to -\infty} \frac{1}{n}\left[F^n - \mathrm{Id}\right].$$

ii) From the last set of inequalities appearing in the proof of proposition 4.2 the following important inequalities may be deduced:

$$\alpha_q = \inf_{\mathbb{R}}(F^q - \mathrm{Id}) \leqslant q\rho \leqslant \sup_{\mathbb{R}}(F^q - \mathrm{Id}) = \beta_q,$$

valid for every integer q.

This shows in particular that for every integer q there exist infinitely many t such that $F^q(t) = t + q\rho(F)$.

iii) If $G = F + r$, $r \in \mathbb{Z}$, we have $\rho(G) = \rho(F) + r$. Hence the map $\rho : \mathcal{K} \to \mathbb{R}$ determines a map, also denoted by ρ, of $\mathcal{H}$ into S^1. We then call $\rho(f)$ the <u>rotation number</u> (of Poincaré) of the homeomorphism f of S^1. (This name will be justified in the next section.)

4.8. <u>THEOREM</u>. The map $\rho : \mathcal{H} \to S^1$ is continuous.

<u>Proof</u>. The map $\rho : \mathcal{K} \to \mathbb{R}$ actually is continuous, for it follows from remark ii) of 4.7 that ρ is the uniform limit of the continuous maps $F \mapsto \frac{1}{n}[F^n - Id]$, $n > 0$, of $\mathcal{K}$ into $\mathbb{R}$. (In view of lemma 4.3 we have

$$\|\rho(F) - \tfrac{1}{n}[F^n - Id]\| \leq \tfrac{1}{n}.)$$

Q.E.D.

4.9. <u>THEOREM</u>. A homeomorphism f of $\mathcal{H}$ has a periodic point of period m if and only if $\rho(F)$ is of the form $p(\frac{n}{m})$, $n \in \mathbb{Z}$.

The homeomorphisms with an "irrational" rotation number are thus ergodic or exceptional. In particular we have another verification that the diffeomorphism constructed in exercise 3.10 is exceptional (cf. example ii) of 4.5).

<u>Proof</u>. Let F be a lift of f. If f has a periodic point of period m then there exist a real number t and an integer n satisfying $F^m(t) = t + n$; thus $F^{rn}(t) = t + rn$ for $r \in \mathbb{Z}$. Letting $q = rm + s$, $0 \leq s < m$ we find

$$\frac{F^q(t)}{q} = \frac{F^s(t)}{q} + \frac{rn}{rm + s} \quad \text{and therefore} \quad \lim_{q \to +\infty} \frac{F^q(t)}{q} = \frac{n}{m}.$$

Conversely, let $\rho(F) = \frac{n}{m}$, $m \in N$ and $n \in \mathbb{Z}$. Then remark ii) of 4.7 entails the existence of a real number t such that

$F^m(t) = t + n$. Hence f has a periodic point of period m. Q.E.D.

4.10. THEOREM. Let f and g denote two homeomorphisms in $\mathcal{H}$. If there exists a continuous map $h: S^1 \to S^1$ of degree 1 and such that $g \circ h = h \circ f$ then $\rho(g) = \rho(f)$. (f is then called a semi-conjugation from f to g.)

Proof. Let F and H be lifts of f and h. Then there is a lift G of g satisfying $G \circ H = H \circ F$. Hence $G^n \circ H = H \circ F^n$ for every integer n, and

$$\frac{G^n(H(t))}{n} = \frac{H(F^n(t))}{n} = \frac{H(F^n(t)) - F^n(t)}{n} + \frac{F^n(t)}{n}.$$

The map h being of degree 1 the function $H - Id$ is periodic of period 1, and we have

$$\rho(G) = \lim_{n \to +\infty} \frac{G^n(H(t))}{n} = \lim_{n \to +\infty} \frac{F^n(t)}{n} = \rho(F).$$ Q.E.D.

4.11. Exercise. The rotation number of the exceptional homeomorphism of S^1 constructed in the proof of proposition 3.8 equals $p(\alpha)$.

4.12. COROLLARY (Topological Invariance of the Rotation Number). Let f and g denote two homeomorphisms of $\mathcal{H}$ which are conjugate in $\mathcal{H}$. Then $\rho(f) = \rho(g)$.

On the other hand, if f and g are conjugate by a homeomorphism reversing the orientation, then we have $\rho(f) = -\rho(g)$.

4.13. PROPOSITION. Let f and g be two commuting homeomorphisms of $\mathcal{H}$. Then $\rho(f \circ g) = \rho(f) + \rho(g)$.

Proof. Let F (respectively G) denote the lift of f (respectively g) such that F(0) (respectively G(0)) belongs to the interval $[0,1)$. Since f and g commute the difference $G \circ F - F \circ G$ is an integer, and we will first show that it vanishes, i.e. F and G commute as well. (Then this will hold for all lifts of f and g.)

If f has a fixpoint we may assume $F(0) = 0$, by conjugating f and g, if necessary, by a common rotation (cf. corollary 4.12). But then $g(p(0))$ is another fixpoint of f, and hence $F(G(0)) = G(0)$ and $G(F(0)) - F(G(0)) = 0$.

If f and g have no fixpoint then we have $|F(t) - G(t)| < 1$ for all $t \in \mathbb{R}$. Then assuming F(0) less than G(0) (i.e. $0 < F(0) \leqslant G(0) < 1$) we deduce the inequalities

$$G(0) < G(F(0)) \leqslant G(G(0)) \quad \text{and}$$

$$F(G(0)) < F(1) = F(0) + 1 \ , \quad \text{and thus}$$

$$-1 + G(0) - F(0) < G(F(0)) - F(G(0)) \leqslant G(G(0)) - F(G(0)) < 1 \ .$$

Hence, again we find $G(F(0)) - F(G(0)) = 0$.

For commuting F and G this leads to

$$\rho(F \circ G) = \lim_{n \to +\infty} \frac{1}{n} (F \circ G)^n(0) = \lim_{n \to +\infty} \frac{1}{n} F^n(G^n(0))$$

$$= \lim_{n \to +\infty} \frac{1}{n} \left[F^n(G^n(0)) - G^n(0)\right] + \lim_{n \to +\infty} \frac{1}{n} G^n(0)$$

$$= \rho(F) + \rho(G) \ .$$

Q.E.D.

Let μ be a Radon measure on S^1. Since the space $\mathcal{P}$ of continuous functions of period 1 on $\mathbb{R}$ is isomorphic to the space of continuous functions on S^1, μ may be identified with a linear and continuous functional on $\mathcal{P}$, and the invariance of μ under a homeomorphism $f = \pi(F)$ of $\mathcal{H}$ corresponds to the invariance under F.

We can then state

4.14. PROPOSITION. Let μ denote a probability measure on S^1 which is invariant under a homeomorphism $f = \pi(F)$ of $\mathcal{H}$. Then $\rho(F) = \mu(F - Id)$.

For $F = Id + \Phi$ we have indeed

$$F^n - Id = \Phi + \Phi \circ F + \dots + \Phi \circ F^{n-1}$$

and therefore $\rho(F) = \mu(\rho(F)) = \mu(\lim_{n\to+\infty} \frac{1}{n}(F^n - Id)) = \mu(\Phi)$.

The following lemma emphasizes the importance of proposition 4.14:

4.15. LEMMA. For every homeomorphism f of S^1 there exists a probability measure on S^1 which is invariant under f.

(This result is a special case of the theorem of Markov-Kakutani [14].)

Proof. Let ν denote Lebesgue measure on S^1. For every positive integer n the expression

$$\mu_n = \frac{1}{n}(\nu + f_*\nu + \dots + f_*^{n-1}\nu)$$

is a probability measure on S^1. Since the set of probability measures on S^1 is countably compact and metrisable (cf. [14]), there is a sequence n_q tending with q toward $+\infty$ such that the sequence μ_{n_q} converges toward a probability measure μ. Then we have

$$f_*\mu = f_*(\lim \mu_{n_q}) = \lim (f_*\mu_{n_q}) = \lim \frac{1}{n_q}(f_*\nu + \dots + f_*^{n_q}\nu) =$$

$$= \lim \frac{1}{n_q}(\nu + f_*\nu + \dots + f_*^{n_q-1}\nu) + \lim \frac{1}{n_q}(f_*^{n_q}\nu - \nu) = \mu$$

Q.E.D.

4.16. Remark. The support of a measure μ on S^1 which is invariant under a homeomorphism f is a closed subspace invariant under f. If, in particular, f has no periodic point, then this support coincides with the unique minimal set M of f (cf. corollary 3.3): In the case where f is an exceptional homeomorphism of S^1 it indeed acts freely on the component set of the open set $U = S^1 - M$, and hence μ vanishes on U.

4.17. Exercise. For every number $\alpha \in \mathbb{R}$ we let $\mathcal{K}_\alpha$ be the closed subspace $\rho^{-1}(\alpha)$ of $\mathcal{K}$. For a rational $\alpha = \frac{n}{m}$, $m \in N$, and a translation $R_n : t \mapsto t + n$, the function $F^m - R_n$ vanishes for every $F \in \mathcal{K}_\alpha$ (cf. remark ii) of 4.7).

Under these circumstances we have

i) the subspaces

$$\mathcal{K}^+_{n/m} = \{F \in \mathcal{K}_{n/m} \mid F^m - R_n \geq 0\} \quad \text{and}$$
$$\mathcal{K}^-_{n/m} = \{F \in \mathcal{K}_{n/m} \mid F^m - R_n \leq 0\} \quad \text{are closed in } \mathcal{K};$$

ii) the complement $U_{n/m}$ of $\mathcal{K}^+_{n/m} \cup \mathcal{K}^-_{n/m}$ with respect to $\mathcal{K}_{n/m}$ is a non-empty open set in $\mathcal{K}$;

iii) the interior of $\mathcal{K}_{n/m}$ is $U_{n/m}$.

On the other hand the interior of $\mathcal{K}_\alpha$ for an irrational α is empty (cf. iii) of exercise 4.18).

4.18. Exercise. For a homeomorphism F of $\mathcal{K}$ we are going to study the function $h : \alpha \mapsto \rho(R_\alpha \circ F)$, where $R_\alpha : t \mapsto t + \alpha$ is a translation.

i) The function h is continuous, increasing, and satisfies $h(\alpha+1) = h(\alpha) + 1$.

ii) Let the homeomorphism F be conjugate in $\mathcal{K}$ to a translation (this is e.g. so for F of class C^2 and $\rho(F)$ irrational, by corollary 5.3). Then the relation $h(\alpha) = h(0) = \rho(F)$ implies $\alpha = 0$. Note that the relation $\rho(G) = \rho(F)$ then implies that the homeomorphism $G \circ F^{-1}$ has a fixpoint.

iii) For F of class C^2 we conclude that the function h assumes exactly once every irrational value.

It also follows that the homeomorphisms having a periodic point are dense in $\mathcal{H}$. (The diffeomorphisms of the form $F = \mathrm{Id} + \Phi$ with Φ a trigonometric polynomial are dense in $\mathcal{H}$.)

iv) If F belongs to $\mathcal{H}^+_{n/m}$ (respectively $\mathcal{H}^-_{n/m}$) and $F^m \neq R_n$, then $h(\alpha) \geqslant \frac{n}{m}$ for $\alpha > 0$ (respectively $h(\alpha) \leqslant \frac{n}{m}$ for $\alpha < 0$), and $h(\alpha) = \frac{n}{m}$ for $\alpha < 0$ (respectively $\alpha > 0$) close to 0.

v) Let $(R_\alpha \circ F)^m \neq R_n$ for every number α and every rational number $\frac{n}{m}$. Then the interval $h^{-1}(\frac{r}{s})$ has a non-empty interior for every rational number r/s.

If, in addition, F does not belong to the boundary of one of the subspaces $\mathcal{H}_\alpha$, $\alpha \in \mathbb{Q}$, then the complement in $[0,1]$ of the interior of $h^{-1}(\mathbb{Q}) \cap [0,1]$ is a compact, perfect, and totally disconnected set (homeomorphic to a Cantor space).

vi) Let $\Phi = F - \mathrm{Id}$ be the restriciton to $\mathbb{R}$ of a holomorphic function on $\mathbb{C}$. (A typical example is $\Phi = a \sin 2\pi t$ with $|a| < \frac{1}{2\pi}$.) If there are integers m and n satisfying $F^m = R_n$, then Φ is a constant. (Note that an analytic automorphism of $\mathbb{C}$ is an affine map $z \mapsto az + b$.)

This result thus leads to the construction of homeomorphisms in $\mathcal{H}$ satisfying the conditions listed under question v).

5. CONJUGATIONS OF CIRCLE HOMEOMORPHISMS TO ROTATIONS

For every number $\alpha \in \mathbb{R}$ we denote by R_α the translation $t \mapsto t+\alpha$, and by $r_\alpha = \pi(R_\alpha)$ the rotation of S^1 through the angle α.

5.1. THEOREM. An ergodic homeomorphism of S^1 with rotation number α is topologically conjugate to the rotation r_α.

We shall actually prove the following more precise result:

5.2. PROPOSITION. Let f be a homeomorphism in $\mathcal{H}$ with an irrational rotation number α. Then there is a continuous map $h: S^1 \to S^1$ of degree 1 such that $h \circ f = r_\alpha \circ h$. If f is ergodic then h is even a homeomorphism. If f has an exceptional minimal set M then h is constant on everyone of the arcs A_i which make up the open set $S^1 - M$. The map f is injective on the complement of the union of the closed arcs $\overline{A_i}$.

(This result is due to H. Poincaré. The proof given here is due to H. Furstenberg [3]. The original proof is presented as exercise 5.6.)

Proof. Let μ be a probability measure on S^1 which is invariant under f. Since f has no periodic point we may apply theorem 4.9 and remark 4.16. Hence we may define a continuous and increasing map $H: \mathbb{R} \to \mathbb{R}$ as follows:

$$H(t) = \mu(p[0,t]) \quad \text{for} \quad t \in [0,1] \;, \quad \text{and}$$

$$H(t) = H(t-n) + n \quad \text{for} \quad t \in [n,n+1] \;, \quad n \neq 0 \,.$$

This function satisfies the relation $H(t+1) = H(t) + 1$, and determines a continuous map $h: S^1 \to S^1$ of degree 1.

For an ergodic f this map is a homeomorphism of $\mathcal{H}$ because the support of μ is then S^1. If f has an exceptional minimal set M the support of μ is M, and h is constant on each of the arcs A_i of which the open set $S^1 - M$ consists, and it is injective on the complement of the union $\bigcup_i \overline{A_i}$.

Let now F be a lift of f. Since μ is invariant under f we have

$$H(F(t)) - H(F(0)) = H(t) - H(0) = H(t) .$$

Hence h is a semi-conjugation from f to the rotation r_β with $\beta = H(F(0))$, i.e. $h \circ f = r_\beta \circ h$. From theorem 4.10 we conclude $\rho(\beta) = \rho(f)$. Q.E.D.

From this proof we can also deduce the following result: Let f be a homeomorphism of S^1 without periodic point, and μ a probability measure on S^1 which is invariant under f. Then

$$\rho(f) = \mu([x, f(x))) \qquad \text{for every } x \in S^1.$$

This result remains true even if f has a periodic point: the support of μ is then a union of periodic points, and for $\rho(f) = \frac{p}{q}$ we have $q\mu([x, f(x))) = p$.

5.3. <u>COROLLARY</u>. Every diffeomorphism of class C^2 of S^1 with an irrational rotation number is topologically conjugate to a rotation.

Indeed, by Denjoy's theorem, f is then ergodic.

5.4. <u>COROLLARY</u>. A direction field of class C^2 on T^2 without a periodic orbit is topologically conjugate to the direction field of a suspension field belonging to a rotation of the circle S^1 through an irrational angle.

5.5 <u>Remarks</u>.

i) A homeomorphism F of $\mathbb{R}$ without fixpoints is conjugate to the translation R_1 for $F(0) > 0$, and to the translation R_{-1} for $F(0) < 0$. The conjugation is effected by an increasing homeomorphism in the same differentiability class as F. (The space of orbits of F is homeomorphic to S^1.)

We are interested here, of course, in conjugation in $\mathcal{K}$ of elements of $\mathcal{K}$, for it corresponds to conjugation of homeomorphisms of S^1 (by orientation-preserving homeomorphisms).

ii) Let f be a homeomorphism of $\mathcal{H}$ which is conjugate in $\mathcal{H}$ to the rotation r_α, and let h and k be two conjugations from f to r_α. Then the homeomorphism $k \circ h^{-1}$ commutes with r_α. For an irrational α it is thus a rotation.

Hence the differentiability class of a conjugation of a homeomorphism of $\mathcal{H}$ to a rotation with an irrational angle is well determined. Even if f is an analytic diffeomorphism such a conjugation may or may not be of class C^1 (cf. exercise 5.10).

5.6. <u>Exercise</u>. Let F be a homeomorphism of $\mathcal{K}$ such that $\alpha = \rho(F)$ is irrational.

i) For any two points x and y of $\mathbb{R}$ the correspondence

$$F^m(x) + n \longmapsto F^m(y) + n$$

(with two integers m and n) is an increasing bijection.

ii) Let m and n be integers. Then the correspondence

$$F^m(x) + n \longmapsto m\alpha + n$$

is an increasing bijection for every $x \in \mathbb{R}$: choose $x = 0$ and note that for $F^q(0)$ between two integers r and s, the same holds for $q\alpha$, hence $pr \leqslant F^{pq}(0) \leqslant ps$ for $p \geqslant 1$.

iii) This correspondence may be extended as a continuous and increasing map $H: \mathbb{R} \to \mathbb{R}$ satisfying $H(t+1) = H(t) + 1$ and $H \circ F = F + \alpha$.

If $f = \pi(F)$ is an ergodic homeomorphism of S^1 then the map H is a homeomorphism of $\mathbb{R}$ which relates F to the translation R_α by conjugation.

5.7. <u>PROPOSITION</u>. Let F be a homeomorphism of $\mathcal{K}$ such that $\rho(F) = \frac{n}{m}$

is rational. The necessary and sufficient condition for F to be conjugate in $\mathcal{K}$ to the translation $R_{n/m}$ (i.e. for $f = \pi(F)$ to be conjugate in $\mathcal{H}$ to the rotation $r_{n/m}$) is $F^m = R_n$.

<u>Proof</u>. The necessity is obvious. Conversely, if $F^m = R_n$, then the map $H = \frac{1}{m}(Id + F + \ldots + F^{m-1})$ is a homeomorphism of $\mathcal{K}$ satisfying $H \circ F = H + \frac{n}{m}$.

Q.E.D.

Note that the constructed conjugation has the same smoothness as F.

5.8. <u>Exercise</u>. For a homeomorphism F of $\mathcal{K}$ to be conjugate in $\mathcal{K}$ to the translation R_α, it is necessary and sufficient that there be a sequence n_q of integers tending towards $+\infty$ with q such that the family (F^{n_q}) is equicontinuous. (If α is irrational an exceptional $f = \pi(F)$ acts freely on the set of components of the complement of its minimal set. For a rational α use proposition 5.7.)

5.9. <u>Exercise: The C^1-conjugation Invariant of M. Herman</u> [5].

Let $\mathcal{K}_1$ denote the subgroup of homeomorphisms of class C^1 of $\mathcal{K}$, and let DF be the derivative of an element F of $\mathcal{K}_1$; it is an element of $\mathcal{P}$.

The metric $d_1(F,G) = \|F - G\| + \|DF - DG\|$ defines a topology on $\mathcal{K}_1$ which is finer than the relative topology induced by $\mathcal{K}$; $\mathcal{K}_1$ thus becomes a topological group.

For $F \in \mathcal{K}_1$ let

$$H_1(F) = \sup_{n \in \mathbb{Z}} \|DF^n\| .$$

i) The map $H_1 : \mathcal{K}_1 \longrightarrow \mathbb{R} \cup \{+\infty\}$ is lower semi-continuous.

ii) For an element H of $\mathcal{K}_1$ the map $t \mapsto D(H^{-1} \circ R_t \circ H)$ is continuous and has period 1. If a diffeomorphism F of $\mathcal{K}_1$ is therefore conjugate in $\mathcal{K}_1$ to a translation, then $H_1(F) < +\infty$.

iii) Conversely we shall show that a diffeomorphism F of $\mathcal{K}_1$ is conjugate in $\mathcal{K}_1$ to the translation R_α, $\alpha = \rho(F)$, if F satisfies $H_1(F) < +\infty$.

Indeed this condition implies by exercise 5.8 that F is conjugate in $\mathcal{K}$ to R_α. If α is rational then by 5.7 F is conjugate to R_α in $\mathcal{K}_1$. Hence we may restrict ourselves to an irrational α and an ergodic $f = \pi(F)$.

iv) $\sup_n \|\mathrm{Log}\, DF^n\| < +\infty$.

Hence by question v) there is a function Φ in $\mathcal{P}$ such that $\mathrm{Log}\, DF = \Phi - \Phi \circ F$ and $\int_0^1 \exp \Phi(t)\, dt = 1$. The function $H: t \longmapsto \int_0^t \exp \Phi(s)\, ds$ is then a conjugation from F to R_α in $\mathcal{K}_1$. Note that $D(H \circ F) = DH$.

v) Let Ψ denote a continuous function on S^1. The following properties are equivalent:

- there is a continuous function φ on S^1 such that $\Psi = \varphi \circ f - \varphi$;
- there is a point τ of S^1 such that

$$\sup_{n \in N} \left| \sum_{i=0}^{n-1} \Psi(f^i(\tau)) \right| < +\infty .$$

(Consider the homeomorphism $k: (\sigma, s) \longmapsto (f(\sigma), s + \Psi(s))$ of $S^1 \times \mathbb{R}$. The ω-limit set of the orbit of the point $(\tau, 0)$ under k contains a compact minimal set M, and M is the graph of a continuous map $\varphi: S^1 \longrightarrow \mathbb{R}$ satisfying $\varphi \circ f - \varphi = \Psi$.)

Note that the above result remains valid if S^1 is replaced by a compact metric space X and f by a homeomorphism of X having all its orbits dense (theorem of Gottschalk-Hedlund).

5.10. <u>Exercise: the examples of V. Arnold</u> [1] (following M. Herman).

Let F be an analytic diffeomorphism of $\mathcal{K}$ of the form

$t \longmapsto t + a \sin 2\pi t + b$, with $0 < |a| < \frac{1}{2\pi}$ and $\rho(F)$ irrational, and let $h: \alpha \longmapsto \rho(R_\alpha \circ F)$ denote the corresponding map introduced in exercise 4.18.

The complement with respect to the interval $[0,1]$ of the interior of the subspace $h^{-1}(\mathbb{Q}) \cap [0,1]$ is then a compact, perfect, and totally disconnected set K.

i) The intersection $D = K \cap h^{-1}(\mathbb{Q})$ is countable and dense in K. For no number $\alpha \in D$ is the diffeomorphism $R_\alpha \circ F$ conjugate to a translation in $\mathcal{K}$.

ii) For every integer $n > 0$ the subspace

$$U_n = \{\alpha \in K \mid H_1(R_\alpha \circ F) > n\}$$

is open in K and contains D.

iii) The subspace $R = \bigcap_{n>0} (U_n - D)$ is dense in $K - D$. The diffeomorphism $R_\alpha \circ F$ is conjugate to a translation in $\mathcal{K}$, but not in $\mathcal{K}_1$. (M. Herman showed in his thesis [5] that a diffeomorphism f of S^1 of class C^r, $3 \leqslant r \leqslant +\infty$ (respectively analytic) is C^{r-2} conjugate (respectively analytically conjugate) to r_α, provided its rotation number α is irrational and a certain Diophantian condition is satisfied.)

iv) On the torus there are analytic direction fields which are topologically conjugate to a constant direction field, but which cannot be defined by a closed form of class C^1 (cf. exercise 1.14).

5.11. <u>Exercise: ergodic uniqueness</u> [3]. Let f be a homeomorphism of S^1 with an irrational rotation number.

i) There is a unique probability measure μ on S^1 which is invariant under f. (If h is a semi-conjugation of f to the rotation r_α then $h_*\mu$ is the Lebesgue measure on S^1. But the support of μ coincides with the minimal set M of f, and h is injective on M except possibly for a countable set.) f is then called uniquely ergodic.

ii) The closure of the subspace $E = \{\varphi \circ f - \varphi\}$ is a hyperplane in the space of continuous functions on S^1.

iii) For every continuous function φ on S^1 the sequence of functions $(\frac{1}{n}\sum_{i=0}^{n-1} \varphi \circ f^i)$ converges uniformly towards $\mu(\varphi)$: for every $\varepsilon > 0$ there exists a continuous function Ψ on S^1 satisfying

$$\|\varphi - \mu(\varphi) - (\Psi \circ f - \Psi)\| < \varepsilon .$$

iv) If f is of class C^1 then $\mu(\mathrm{Log}\, f') = 0$, because $\mathrm{Log}(f^n)' = \sum_{i=0}^{n-1} (\mathrm{Log}\, f') \circ f^i$ and $\int_{S^1} (f^n)'\, d\theta = 1$.

Let therefore f have an exceptional minimal set M, and let I be one of the arcs making up the open set $S^1 - M$. Then the length l_n of the arc $I_n = f^n(I)$, $n \geq 1$, satisfies $\lim_{n\to+\infty} \sqrt[n]{l_n} = 1$.

v) Let F be a lift of f, and $H = \mathrm{Id} + \Phi$ a conjugation in $\mathcal{K}$ from F to the translation R_α such that $\mu(\Phi) = 0$. Then the sequence of homeomorphisms $S_n = \frac{1}{n}\sum_{q=0}^{n-1} (F^q - q\alpha)$, $n > 0$, of $\mathcal{K}$ converges uniformly towards H. (Note that

$$F^q - \mathrm{Id} - q\alpha = \Phi - \Phi \circ F^q \quad \text{and} \quad H - S_n = \frac{1}{n}\sum_{q=0}^{n-1} \Phi \circ F^q .)$$

APPENDIX A: HOMEOMORPHISM GROUPS OF AN INTERVAL

In view of applications to foliations we shall study in the appendices A and B the possibilities of generalising Denjoy's theorem (cf. theorem 3.4) to finitely generated groups of homeomorphisms of the interval $[0,1]$ and of the circle S^1. As for direction fields on surfaces we classify the orbits of these groups as proper, locally dense, and exceptional (cf. I-2.8).

An Abelian finitely generated group of increasing homeomorphisms of the interval $[0,1]$ is free, and each of its orbits has a stability subgroup which is a direct summand. If this group is isomorphic to $\mathbb{Z}$ then all of its orbits are proper and have a fixed point of the group as α-limit and ω-limit sets.

A.1 THEOREM. Let G be an Abelian group of increasing C^2-diffeomorphisms of the interval $[0,1]$ of finite rank at least equal to 2, and let 0 and 1 be the only fixpoints of G. Then all its orbits in the open set $(0,1)$ are everywhere dense.

Proof. We first show that there is an element of G which has no fixpoint in $(0,1)$: assume that all the elements $f_1, \ldots, f_r$ of a base of G have a fixpoint in $(0,1)$, and denote by x_1 a fixpoint of f_1 in $(0,1)$. Then one of the sequences $(f_2^n(x_1))$ and $(f_2^{-n}(x_1))$, $n \geq 0$, has a point $x_2 \in (0,1)$ as limit which remains fixed under both f_1 and f_2. By iteration we obtain a point $x_r \in (0,1)$ which is fixed by G, contrary to the assumptions.

Let now f be an element of G without fixpoint in $(0,1)$ which we may assume to be a contraction on $[0,1)$. By corollary II-A.4 and theorem II-A.7 there exists then an increasing homeomorphism of $(0,1)$ onto $\mathbb{R}$ carrying G by conjugation into a group of translations of $\mathbb{R}$ whose rank is at least 2. Hence each of the orbits of G in $(0,1)$ is everywhere dense. Q.E.D.

A.2. Remarks.

i) If G is an Abelian group of rank 2 acting freely on $(0,1)$ (i.e. all elements different from the identity have no fixpoint in $(0,1)$)

then theorem A.1 is a consequence of Denjoy's theorem. Let indeed $\{f,g\}$ be a base of G with f a contraction on $[0,1]$. Then we may construct a C^2-diffeomorphism of $(0,1)$ onto $\mathbb{R}$ which transforms f into the translation $t \mapsto t+1$ by conjugation, and g into an increasing C^2-diffeomorphism h satisfying $h(t+1) = h(t) + 1$. Then h has an irrational rotation number because of our assumptions about G.

ii) The example constructed in remark ii) of II-A.11 yields an Abelian group of rank 2 of increasing homeomorphisms of $[0,1]$ all of whose orbits are proper.

iii) An increasing homeomorphism of $\mathbb{R}$ onto the interval $(0,1)$ may be used to transform a group of increasing homeomorphisms of $\mathbb{R}$ into a group of increasing homeomorphisms of $[0,1]$.

Starting with an exceptional homeomorphism of the circle S^1 we may thus construct an Abelian group of rank 2 of increasing homeomorphisms of $[0,1]$ having proper and exceptional orbits.

A.3. PROPOSITION. Let G be a group of increasing homeomorphisms of the interval $[0,1]$ onto itself. If it acts freely on $(0,1)$ then it is Abelian.

Proof. The relation "$f(t) \leq g(t)$ for every $t \in [0,1]$" is an order on G which is compatible with the group structure. This order is total and Archimedean: the contractions on $[0,1)$ are the elements less than the identity, and if f and g are two of these contractions and $t \in (0,1)$ then there is an integer $n > 0$ such that $g^n(t) \leq f(t)$, thus $g^n \leq f$.

The proof is concluded by the next lemma. Q.E.D.

A.4. LEMMA. An ordered group with a total and Archimedean order is Abelian.

Proof. A totally ordered group is generated by the set P of its elements which are greater than its neutral element. If P has a smallest element then P is therefore cyclic.

Assume now that P has no smallest element. For every $x \in P$ there is then a $y \in P$ such that $y^2 \leqslant x$ (if $z < x < z^2$, then $(xz^{-1})^2 < x$). For two elements x and y of an Archimedean G satisfying $x < y$ there exists therefore $z \in P$ and an integer $n \geqslant 2$ such that $x \leqslant z^n < z^{n+1} \leqslant y$: choose $z < x$ with $z^2 \leqslant yx^{-1}$.

Let now a be a fixed element of P. For every $x \in P$ the set of all rational numbers $\frac{p}{q}$ with positive p and q and such that $x^q \leqslant a^p$, is a non-empty interval of $\mathbb{Q}$. It is bounded from below by $\frac{1}{n}$ if $x^{n-1} \leqslant a < x^n$, $n \geqslant 1$. Hence we may define a map $h: P \to (0,+\infty)$ by $h(x) = \inf(\frac{p}{q} > 0 \mid x^q \leqslant a^p)$. This map is increasing and injective: from $h(x^n) = nh(x)$ we see that $x \leqslant z^n < z^{n+1}$ implies

$$h(x) \leqslant nh(z) < (n+1)h(z) \leqslant h(y) .$$

The proof will be established if we can show that $h(xy) = h(x) + h(y)$.

But for every $\varepsilon > 0$ which is less than both $h(x)$ and $h(y)$ there is $z \in P$ satisfying $h(z) < \varepsilon$. The inequalities $z^p \leqslant x < z^{p+1}$ and $z^q \leqslant y < z^{q+1}$ with positive p and q then imply

$$ph(z) \leqslant h(x) < (p+1)h(z) ,$$

$$qh(z) \leqslant h(y) < (q+1)h(z) , \quad \text{and}$$

$$(p+q)h(z) \leqslant h(xy) < (p+q+z)h(z) , \quad \text{hence}$$

$$|h(xy) - h(x) - h(y)| \leqslant 2\varepsilon . \qquad \text{Q.E.D.}$$

A.5. Remark. We have actually shown in this proof that h may be extended to an increasing isomorphism of G onto a subgroup of $\mathbb{R}$, and that two such increasing isomorphisms differ by a constant positive factor.

In particular we may deduce that two groups K and K' of increasing homotheties of $\mathbb{R}$ are C^0-conjugate if and only if there is an isomorphism $\varphi: K \longrightarrow K'$ and a number $\mu > 0$ such that $\varphi(k)$ is the homothety of ratio λ^μ for every homothety $k \in K$ of ratio λ.

A.6. Exercise. Let G_r denote the group of germs at 0 of increasing local C^r-diffeomorphisms of $\mathbb{R}$ leaving the origin fixed, $0 \leqslant r \leqslant +\infty$. Let G_r be provided with the partial order induced by the comparison of graphs near 0. Moreover let H denote a finitely generated subgroup of G_r, $r \geqslant 2$, which has a total Archimedean order.

i) H is conjugate in G_0 to a group of germs of homotheties of $\mathbb{R}$ (cf. remark 1) of A.2).

ii) Let no element of H different from the identity be tangent to the identity. Then H is C^1-conjugate (actually C^{r-1}-conjugate) to the group of its jets of order 1 (cf. theorem II-A.3).

iii) Let all elements of H be tangent to the identity of order k, $1 \leqslant k < r$. Then H is C^0-conjugate to the group of homotheties $t \longmapsto e^a t$, where a runs through the set of $(k+1)^{st}$ derivatives at 0 of the elements of H (cf. exercise II-A.12).

APPENDIX B: HOMEOMORPHISM GROUPS OF THE CIRCLE

B.1. THEOREM. An Abelian and finitely generated group of C^2-diffeomorphisms of the circle S^1 has no exceptional orbit.

If such a group contains a diffeomorphism without periodic point then all its orbits are dense by Denjoy's theorem. Hence we may assume that every element of G has a periodic point. The two following lemmas then show that the above result is an immediate consequence of theorem A.1 (or of Denjoy's theorem).

B.2. LEMMA. Let an Abelian group H of homeomorphisms of S^1 have a finite system of generators each of which has a periodic point. Then there is a subgroup of finite index in H having a fixpoint.

Hence every element of H has a periodic point (see also theorem 4.9 and proposition 4.13).

Proof. Let $(f_1,\ldots,f_n)$ be a system of generators of H where each f_i has a periodic point (of order q_i), and let K denote the subgroup of finite index in H generated by $g_1 = f_1^{q_1},\ldots,g_n = f_n^{q_n}$. If u_1 is a fixpoint of g_1 then the points $g_2^k(u_1)$, $k > 0$, are fixpoints of g_1 as well. They have as limit a point u_2 which remains fixed under both g_1 and g_2. By iteration of this procedure we obtain a point u_n which remains fixed under all $g_1, g_2, \ldots$ and g_n, hence under K. Q.E.D.

B.3. LEMMA. Let H be a group of homeomorphisms of a manifold M, and K a subgroup of finite index in H. Then the orbit of a point $x \in M$ under H is proper (respectively locally dense) if and only if the same holds for the orbit of x under K.

Proof. If the orbit of x under H (respectively K) is proper (respectively locally dense) then the same statement is true for the orbit of x under K (respectively H).

Conversely, by the assumptions about K, the orbit γ of x under H is a finite union $\gamma_1 \cup \ldots \cup \gamma_n$ of subspaces of M which are permuted under the action of H and such that $\gamma_1 = Kx$ is the orbit of x under K. Hence γ is proper if γ_1 is, and γ_1 is locally dense if γ is (Baire property). Q.E.D.

B.4. COROLLARY. Let G be an Abelian and finitely generated group of C^2-diffeomorphisms of the circle S^1. Then a minimal set of G is finite if it does not coincide with S^1.

B.5. Remark. Let G be an Abelian group of diffeomorphisms of S^1 which preserve the orientation, and let $g \in G$ be ergodic (in particular g of class C^r, $r \geqslant 2$, and without a periodic point). Then G is topologically conjugate to a rotation group: by theorem 5.1 g is topologically conjugate to a rotation through an irrational angle, and every homeomorphism which preserves the orientation and commutes with such a rotation is itself a rotation.

B.6. PROPOSITION. Let M be a closed, perfect, and totally disconnected subspace of S^1. For every integer $n \geqslant 2$ there exist n exceptional homeomorphisms of S^1 which commute mutually and have M as minimal set.

Proof. Let $\alpha_1, \ldots, \alpha_n$ denote n irrational numbers which are independent over the rationals, and let γ be the following subset of S^1:

$$\gamma = p(\sum_{i=1}^{n} r_i \alpha_i), \quad r_i \in \mathbb{Z} .$$

There is a bijection λ of γ onto the set Γ of components of the open set $U = S^1 - M$ which is compatible with "the order on S^1". We set

$$I_{r_1, \ldots r_n} = \lambda(p(\sum_{i=1}^{n} r_i \alpha_i)) .$$

As in 2.6 we now construct a homeomorphism f_1 of S^1 having M as minimal set and such that

$$f_1(I_{r_1,r_2,\ldots,r_n}) = I_{r_1+1,r_2,\ldots,r_n} .$$

Assume that homeomorphisms $f_1,\ldots,f_{k-1}$ have been found such that

$$f_i(I_{r_1,\ldots,r_i,\ldots,r_n}) = I_{r_1,\ldots,r_i+1,\ldots,r_n} ,$$

and such that any two of them commute. Then we construct a homeomorphism f_k as follows:

On $I_{0,\ldots,0,r_k,\ldots,r_n}$ f_k is the increasing affine isomorphism onto $I_{0,\ldots,0,r_k+1,\ldots,r_n}$;

on $I_{r_1,\ldots,r_k,\ldots,r_n}$ we define

$$f_k = f_{k-1}^{r_{k-1}} \circ \ldots \circ f_1^{r_1} \circ f_k \circ f_1^{-r_1} \circ \ldots \circ f_{k-1}^{-r_{k-1}} .$$

This homeomorphism commutes with $f_1,\ldots,f_{k-1}$, and M is its minimal set. Q.E.D.

B.7. <u>THEOREM</u>. There is a group of analytic diffeomorphisms of S^1 having two generators and an exceptional minimal set.

<u>Proof</u>. We shall construct G as a subgroup of the automorphism group A of the complex projective line (i.e. of the sphere S^2). (The group A is described e.g. in the treatise of L. Ford on automorphic functions). G will be generated in A by two transformations f and g defined as follows:

- f is the composition of the inversion with respect to the circle of radius 2 and centre 1 and the axial symmetry with respect to the imaginary axis;

- g is the composition of the inversion in the circle of radius 1 and centre 5 and the axial symmetry with respect to the line x=5.

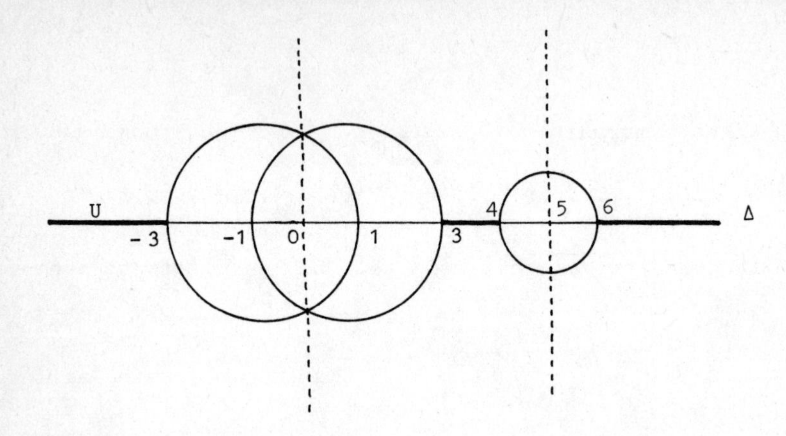

Both these maps are elliptic and have finite orders 3 and 2; their analytic expressions are

$$f(z) = \frac{3+z}{1-z}, \qquad g(z) = \frac{26-5z}{5-z}.$$

Since they both leave the real axis Δ invariant, G may be considered as a group of analytic diffeomorphisms of the circle Γ of $\mathbb{P}\mathbb{C}^1$ whose projection onto $\mathbb{C}$ is Δ.

The transforms of the open set in Γ

$$U = \Gamma - ([-3,+3] \cup [4,6])$$

by the elements of G are mutually disjoint: the sets $f^n(U)$, $n = 1,2$ (respectively $g(U)$) are indeed contained in the interval $I = [-3,+3]$ (respectively $J = [4,6]$), and we have $g(I) \subset J$ and $f^n(J) \subset I$, $n = 1,2$.

Hence the complement with respect to Γ of the saturated set of U under G is a non-empty, compact, and invariant set. Thus it contains a minimal set $M \neq \Gamma$. M cannot be finite because the transformations $f \circ g$ and $g \circ f$ are both hyperbolic with different fixpoints, and the limit sets of the orbits of a hyperbolic map are its two fixpoints. Hence M is an exceptional minimal set. Q.E.D.

B.8. Remarks.

i) The group G has actually a single minimal set, since such a minimal set is infinite and contains the fixpoints of all hyperbolic transformations of G.

ii) Since any two images of U under maps of G are disjoint the group G is isomorphic to the free product of the cyclic groups $\mathbb{Z}/3\mathbb{Z}$ and $\mathbb{Z}/2\mathbb{Z}$.

More generally one may construct a group of analytic diffeomorphisms of S^1 with an exceptional minimal set which is isomorphic to the free product of a finite family of cyclic groups (finite or infinite).

iii) The automorphism of $\mathbb{P}\mathbb{C}^1$

$$h: z \longmapsto \frac{z - i\sqrt{3}}{z + i\sqrt{3}}$$

transforms Γ into the unit circle S^1 of $\mathbb{C}$, and it is a conjugation from f to the rotation through $2\pi/3$ of S^1.

B.9. PROPOSITION. A homeomorphism group of S^1 which acts freely on S^1 is Abelian.

Indeed $\pi^{-1}(G)$ is a subgroup of K which carries a total and Archimedean order.

Under these circumstances the map ρ is an injective homeomorphism of G into S^1 (cf. proposition 4.13 and remark ii) of 4.7). By remark B.5 such a group of homeomorphisms is topologically conjugate to a rotation group of S^1 if it is finite or if it contains an ergodic element.

B.10. PROPOSITION. An infinite homeomorphism group of S^1 which acts freely on S^1 has a single minimal set.

As in the proof of proposition 3.1 we indeed recognise that this minimal set is the limit set $\overline{\gamma - \gamma}$ of each orbit γ of G. It either coincides with S^1 or it is a compact, perfect, and totally disconnected set.

B.11. THEOREM. Let G be an infinite group of homeomorphisms of S^1 acting freely on S^1, and let the minimal set of G coincide with S^1. Then G is topologically conjugate to a rotation group.

Proof. Since S^1 is minimal G does not contain exceptional homeomorphisms (otherwise their minimal set would be invariant under G). By B.5 we may thus assume that every element of G is of finite order. Then there is a sequence $(f_n)_{n>0}$ of elements of G and a point $u \in S^1$ with the following properties:

- the sequence $\rho(f_n)$ tends towards an irrational "number" α;
- the sequence $f_n(u)$ tends towards a point v;
- the map f of the orbit of u into the orbit of v defined by $f(g(u)) = g(v)$ is then compatible with the "order" on S^1. It may be extended therefore as a homeomorphism f of S^1 which is the limit of the sequence (f_n) and which commutes with G.

Finally the group H generated by G and f acts freely on S^1:

if an element $h = gf^n$, $g \in G$, of H has a fixpoint then

$$\rho(h) = \rho(g) + n\alpha = 0 ;$$

hence h is the identity. Q.E.D.

B.12. Exercises (H. Imanishi, J.Math. Kyoto Univ., 14,1974).

i) There is a (non-finitely generated) group of C^∞-diffeomorphisms of S^1 which acts freely on S^1 and has an exceptional minimal set.

ii) A compact subgroup of the group $\mathcal{H}$ of homeomorphisms of S^1 is topologically conjugate to a rotation group.

Chapter V. Vector Fields on Surfaces

1. CLASSIFICATION OF COMPACT SURFACES

1.1. Given an integer $g \geq 0$ we denote by T_g:

- the sphere S^2 if $g = 0$;
- the torus T^2 if $g = 1$;
- the connected sum of g copies of T^2 if $g \geq 2$

(cf. exercise vi) of I-A.2).

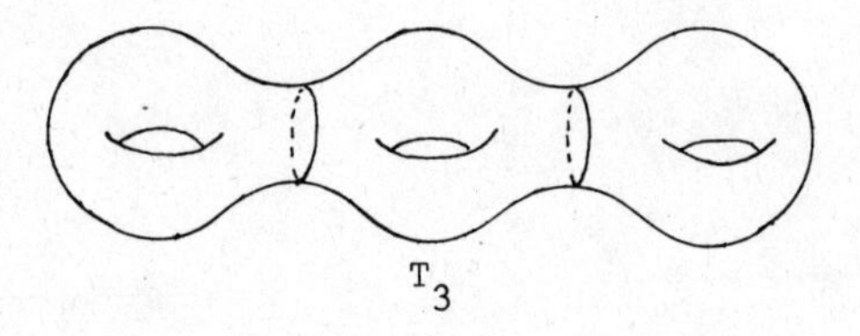

T_3

By P_g we mean

- the real projective plane $\mathbb{P}\mathbb{R}^2$ for $g = 0$;
- the connected sum of $g+1$ copies of $\mathbb{P}\mathbb{R}^2$ for $g \geq 1$. In particular the surface P_1 is diffeomorphic to the Klein bottle K^2 (cf. remark iii) of IV-2.4).

The number g appearing in the compact, connected, and orientable surface without boundary T_g is called its <u>genus</u> (T_g is also called the "surface with g holes"). For the compact, connected, and non-orientable surface without boundary P_g the number g is again called its genus.

1.2. Exercises.

i) The fundamental group of T_g is generated by 2g generators $a_1, b_1, \ldots, a_g, b_g$ satisfying the relation $a_1 b_1 a_1^{-1} b_1^{-1} \ldots a_g b_g a_g^{-1} b_g^{-1} = 1$.

In particular $\Pi_1(T_g)$ has a quotient group which is free on g generators.

ii) The fundamental group of P_g is a group with g generators $c_1, \ldots c_g$ subject to the relation $c_1^2 \ldots c_g^2 = 1$.

1.3. PROPOSITION. The Euler characteristic of T_g equals $2-2g$, and for P_g it is $1-g$.

(Recall that the Euler-Poincaré characteristic $\chi(M)$ of a compact manifold M equals the alternating sum of the dimensions of its de Rham cohomology spaces (cf. [16]).)

Proof. First we have $\chi(T_o) = 2$, $\chi(T_1) = 0$, and $\chi(P_o) = 1$. Let then denote by M and N two compact surfaces without boundary. Removing from each of M and N a point we obtain two surfaces U and V. The exact Mayer-Vietoris sequence in the de Rham cohomology for $M \# N$ reads as follows:

$$0 \longrightarrow H^o(M \# N) \longrightarrow H^o(U) \oplus H^o(V) \longrightarrow H^o(U \cap V) \longrightarrow$$
$$\longrightarrow H^1(M \# N) \longrightarrow H^1(U) \oplus H^1(V) \longrightarrow H^1(U \cap V) \longrightarrow$$
$$\longrightarrow H^2(M \# N) \longrightarrow 0.$$

Since the open set $U \cap V$ is diffeomorphic to the cylinder $S^1 \times \mathbb{R}$ we deduce $\chi(M \# N) = \chi(U) + \chi(V)$. In particular if N is the sphere S^2 then $M \# N$ is diffeomorphic to M and V to $\mathbb{R}^2$, hence $\chi(U) = \chi(M) - 1$, $\chi(M \# N) = \chi(M) + \chi(N) - 2$. By induction we arrive at the above values.

Q.E.D.

From 1.3 we can deduce in particular that any two of the

surfaces T_g (respectively P_g) are non diffeomorphic. Actually we have more precisely (cf. [17]):

1.4. <u>THEOREM</u> (<u>Classification of surfaces</u>). A smooth, compact, connected orientable (respectively non-orientable) surface without boundary is diffeomorphic to one of the surfaces T_g (respectively P_g).

Orientable (respectively non-orientable) surfaces are thus classified by their Euler characteristic which is of the form $2-2g$ (respectively $1-g$) with $g \geqslant 0$. As well we may use their first Betti number b_1 for de Rham cohomology which is even (respectively arbitrary). Their genus equals then $\frac{1}{2}b_1$ (respectively b_1).

1.5. <u>COROLLARY</u>. The orientation covering space of the surface P_g is the surface T_g.

If $p:M \to N$ is indeed a two-sheeted covering of a compact manifold N, then $\chi(M) = 2\chi(N)$.

1.6. <u>Exercises</u>.

i) The surface $T_g \# P_o$ (respectively $T_g \# P_1$) is diffeomorphic to P_{2g} (respectively P_{2g+1}).

ii) The smooth, compact, connected, orientable (respectively non-orientable) surfaces with boundary are classified by their first Betti number for de Rham cohomology and the number p of boundary components. We define the genus of such a surface M as the integer

$$g = \tfrac{1}{2}\,(1+b_1-p) \quad (\text{respectively } g = b_1 - p);$$

then M is diffeomorphic to a submanifold of T_g (respectively P_g).

iii) Let M be a compact surface, and J a Jordan curve in the interior of M. If U is the interior of a tubular neighbourhood of J in M then $N = M - U$ is called the surface obtained from M by cutting M along J. It has one or two components; if the boundary of M has p components then N has $p+1$ or $p+2$ boundary components, according to whether J is one- or two-sided.

If the genus of M vanishes then the same holds for each component of N. If the genus g of M is different from zero then one of the components of N has a genus which is strictly less than g.

iv) Let M be a compact, connected, orientable (respectively non-orientable) surface of genus g, and let K be the union of p mutually disjoint Jordan curves in the interior of M. If $p \geq g+1$ (respectively $g+2$) then the open set $M - K$ is not connected. If $p = g$ (respectively $g+1$) and $U = M - K$ is connected, then U is diffeomorphic to an open set in $\mathbb{R}^2$. (Note that this situation occurs e.g. on the torus T^2 and the Klein bottle K^2).

2. VECTOR FIELDS ON SURFACES

2.1. THEOREM (Hopf). The torus T^2 and the Klein bottle K^2 are the only smooth, compact, connected surfaces without boundary having a vector field without singularity.

Proof. From the theorem of Gauss-Bonnet (theorem A.11) and from proposition A.9 we deduce that a compact connected surface without boundary carrying a vector field without singularity has vanishing Euler characteristic.

Conversely the vector field on $\mathbb{R}^2$ with components $P = 0$ and

$Q = 1$ is invariant under the action of the group G of IV-2.1; hence it induces on K^2 a vector field without singularity. Similarly for T^2.

Q.E.D.

2.2. Remarks.

i) The coordinate z induces on the torus in $\mathbb{R}^3$

$$x^4 + y^4 + z^4 + 2x^2y^2 + 2y^2z^2 + 2z^2x^2 + 6x^2 - 10y^2 - 10z^2 + 9 = 0$$

a smooth function. Its gradient (with respect to the Riemannian structure induced by the Euclidean structure of $\mathbb{R}^3$) has exactly four singular points, viz. two nodes and two saddles.

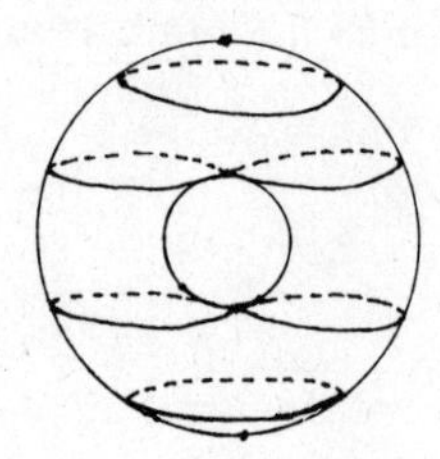

Using this vector field we may construct on the surface T_g, $g \geqslant 2$, a vector field having $2g+2$ singular points, viz. 2 nodes and 2g saddles (cf. exercise 3.3).

ii) Let a vector field be determined on $\mathbb{R}^2$ by the components $P = x$ and $Q = 2y$.

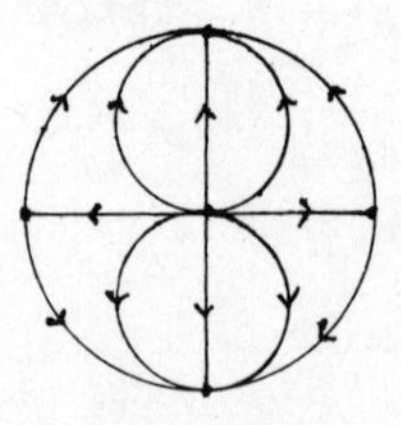

By Poincaré's method of compactification (cf.III-1.9) this vector field induces on the projective plane $\mathbb{P}\mathbb{R}^2$ a vector field having three singular points: 2 nodes and 1 saddle. Starting from this field we can construct on the surface P_g, $g \geq 1$, a vector field with $g+3$ singular points: 2 nodes and $g+1$ saddles.

2.3. Exercise. We shall show that on every connected and non-compact surface without boundary there exists a vector field without singularity.

i) Let V be a compact surface with boundary. Then there is a vector field on V which is transverse to the boundary and has a finite number of singular points: Take e.g. the gradient (with respect to an arbitrary Riemannian structure on V) of a smooth function on V which is constant on ∂V and has a finite number of critical points, all of them in the interior of V.

ii) Let M be a connected and non compact surface without boundary. Then there is a sequence $(V_n)_{n \geq 0}$ of compact submanifolds with boundary with the following properties:

$V_o = \emptyset$;

V_{n+1} lies in the interior of V_n;

$M = \bigcup_n V_n$.

(Choose a smooth, positive, and proper function f on M. By Sard's theorem there is then a positive and increasing sequence $(t_n)_{n \geq 0}$ of regular values which tend towards $+\infty$. Define $V_n = f^{-1}([0,t_n])$.)

iii) There is a vector field X on M whose restriction to V_n is transverse to ∂V_n and has a finite number of singular points, all of which lie in the interior of V_n.

iv) For every $n \geq 1$ there is a diffeomorphism h of M which is the identity map outside of $V_{n+1} - V_{n-1}$ such that the vector field $Y_n = h_n^T \circ X \circ h_n^{-1}$ has no singularity on V^n (cf. theorem I-A.5). The vector field $Y = \lim_{n \to +\infty} Y_n$ has no singularity on M.

v) On every compact surface with (non-empty) boundary there is a vector field without singularity.

vi) The only compact and connected surfaces with (non-empty) boundary on which there is a vector field without singularity and tangent or transverse to each boundary component are the cylinder and the Moebius strip (cf. theorem 3.1).

vii) On every compact and connected surface without boundary there exists a vector field with a single singular point.

2.4. THEOREM (A.J.Schwartz [8]). A vector field of class C^r, $r \geq 2$, on a surface without boundary has no exceptional minimal set.

Thus the theorem of Schwartz generalizes the theorem of Denjoy for the torus to arbitrary surfaces. (It also offers a new proof for it.)

Proof. Let M be a surface without boundary, and X a vector field on M of class C^r, $r \geq 2$, having an exceptional minimal set E. Furthermore, let N denote a compact arc transverse to E which intersects E and has its endpoints outside of E.

The intersection K of E and N is then a compact, perfect and totally disconnected set lying in the interior $\hat{N}$ of N. As in II-7.1 we may define a Poincaré map P on an open neighbourhood U of K in $\hat{N}$: for a point $x \in U$, P(x) is the first point of intersection of the positive half-orbit γ_x^+ with $\hat{N}$. This Poincaré map is a diffeomorphism of class C^r of U onto an open set of $\hat{N}$ for which K is a minimal set (i.e. leaving K invariant and such that every orbit of f in K is dense in K).

The proof is concluded by applying the next lemma.

Q.E.D.

2.5. LEMMA. A local diffeomorphism of class C^r, $r \geq 2$, of $\mathbb{R}$ has no compact exceptional minimal set.

Proof. Let f be a C^r-diffeomorphism, $r \geq 2$, of an open set $V \subset (0,1)$ onto an open set of $(0,1)$ having a compact exceptional minimal set K. There exist then an open neighbourhood W of K in V and two numbers L and M greater than 1 satisfying the following properties:

i) $\overline{W} \subset V$;

ii) $|f'|$ lies between $\frac{1}{L}$ and L on $\overline{W}$;

iii) $|f''| < M$ on $\overline{W}$.

Denote by d the distance from K to $V-W$. Since the intervals which make up $V-K$ are mutually disjoint, and only a finite number of them have length greater than d, there is an interval (a,b) in $W-K$ with endpoints on K and such that for every $n \geq 0$ we have:

f^n is defined on $[a,b]$ and

$f^n([a,b])$ is contained in W.

Then the series $\sum_{n=0}^{+\infty} |Df^n(a)|$ converges: we may indeed find a sequence (s_n) in (a,b) such that

$$\sum_{n=0}^{+\infty} |Df^n(s_n)| \leq \frac{1}{b-a} \sum_{n=0}^{+\infty} |f^n(b) - f^n(a)| \leq \frac{1}{b-a},$$

and a computation similar to the one carried out in the proof of Denjoy's theorem (cf. IV-3.4) shows

$$|Df^n(a)| \leq |Df^n(s_n)|\, e^{LM}.$$

Let σ denote the sum of the above series, and let

$$\delta = d/3LM(\sigma+1), \quad J = [a-\delta, a+\delta].$$

We will now show that the sequence $(Df^n)_{n\geq 0}$ converges uniformly to zero on J. For this purpose we verify by induction the following inequalities:

(a_n) $\quad |f^n(t) - f^n(a)| \leqslant d$ for every $t \in J$;

(b_n) $\quad |Df^n(t)| \leqslant 3|Df^n(a)|$ for every $t \in J$.

They are obviously satisfied for $n = 0$, and (a_n) implies that f^{n+1} is defined on J. Moreover, if (a_k) and (b_k) are valid for $k \leqslant n$ then the same holds for (a_{n+1}) and (b_{n+1}), because of

$$|Df^{n+1}(t)|/|Df^{n+1}(a)| \leqslant \exp\left[LM \sum_{k=0}^{n} |f^k(t) - f^k(a)|\right]$$

$$\leqslant \exp\left[\delta LM \sum_{k=0}^{n} |Df^k(s_k)|\right]$$

$$\leqslant \exp\left[3\delta LM\sigma\right] \leqslant 3;$$

(with $s_k \in J$) and

$$|f^{n+1}(t) - f^{n+1}(a)| \leqslant |t-a| \; |Df^{n+1}(s)| \qquad (s \in J)$$

$$\leqslant 3\delta|Df^{n+1}(a)| \leqslant 3\delta\sigma \leqslant d.$$

Choose now an integer $n > 0$ such that

$$|f^n(a) - a| \leqslant \frac{\delta}{2} \quad \text{and} \quad |Df^n| \leqslant \tfrac{1}{2} \quad \text{on } J.$$

The diffeomorphism f^n is then a contraction which sends J into itself; hence it has in J a unique fixpoint which must belong to K. This is a contradiction. Q.E.D.

2.6. <u>COROLLARY</u>. Let M be a compact surface without boundary, and X a vector field on M of class C^r, $r > 2$. Then a minimal set of X is of one of the following three types:

- a singular point,
- a periodic orbit,
- the whole surface M - in which case M is

diffeomorphic to the torus T^2.

Indeed a non exceptional minimal set which is neither a singular point nor a periodic orbit has interior points; hence it coincides with M. But then M has vanishing Euler characteristic (theorem 2.1) and is diffeomorphic to T^2 by theorem IV-2.5. Q.E.D.

2.7. COROLLARY. Let X be a vector field of class C^r, $r \geq 2$, on a surface without boundary M, and let γ denote an ω-stable (respectively α-stable) orbit of X in Lagrange's sense. If the limit set Ω_γ (respectively A_γ) does not coincide with M and does not contain a singular point then it reduces to a periodic orbit.

For a sufficiently smooth vector field this result generalizes the theorem of Poincaré-Bendixson (corollary III-1.6) to the case of an arbitrary surface.

Its proof is also comparable: the assumptions about X and γ imply that Ω_γ (respectively A_γ) is a non-empty compact set which contains a periodic orbit (corollary 2.6). By proposition II-7.11 Ω_γ (respectively A_γ) then reduces to this periodic orbit. Q.E.D.

2.8. Remarks.

i) R. Sacksteder and A.J.Schwartz have shown that for class C^1 and under the assumptions of corollary 2.7 the set Ω_γ is always a minimal set (Ann.Inst.Fourier, 15,2,1965, pp.201-214).

ii) On an open set of the sphere S^2 or of the real projective plane $\mathbb{PR}^2$ a C^r-vector field with $r \geq 1$ has no exceptional orbit (proposition III-1.1).

2.9. Exercises.

i) Let M be a surface with boundary of genus g. Then a C^1 vector field on M which is tangential on the boundary has at most 2g-1 exceptional minimal sets.

ii) Let $\Phi: \mathbb{R}^2 \times M \to M$ be an action of class C^2 (actually C^1 would be sufficient) of $\mathbb{R}^2$ on the compact surface M which leaves invariant every boundary component of M. If the Euler characteristic $\chi(M)$ does not vanish then Φ has a fixpoint (cf. E.Lima: Comm.Math.Helv., 39, 1964, pp.97-110).

In other words: Two C^2 vector fields X and Y on M which are tangential on the boundary have a common zero if their bracket vanishes. The proof is carried out by induction on the genus of M; if Φ had no fixpoint it would have an uncountable family of minimal sets which would all be diffeomorphic to S^1, and there would exist one which would not bound a cylinder with any boundary component of M.

3. THE INDEX THEOREM

By proposition III-4.20 the index of a vector field on an open set of $\mathbb{R}^2$ at an isolated singular point is invariant under C^1-conjugation. This makes it possible to define the index of a vector field X on a surface M at an isolated singular point as the index of the vector field's "expression" in an arbitrary local chart.

We then have

3.1. THEOREM (Index theorem of Hopf). Let X be a smooth vector field on a compact surface without boundary M, and let X have a finite number of singular points. Then the sum of the indices of X at the singular points equals the Euler characteristic of M.

Proof. If M is not orientable the Euler characteristic of the orientation covering space $\hat{M}$ of M is twice the Euler characteristic of M. If moreover $\hat{X}$ is the lift of the vector field X then $\hat{X}$ has twice the number of singular points of X, and their indices do not change under projection. We may therefore assume that M is oriented.

Let then denote the singular points of X by $u_1,\ldots,u_n$, and let $D_1,\ldots,D_n$ be compact and mutually disjoint submanifolds of M, each diffeomorphic to the disc D^2, and such that D_i contains u_i in its interior. Then a submanifold V is determined as follows: its interior is $M-(D_1\cup\ldots\cup D_n)$, and its boundary is the union of the Jordan curves $C_i=\partial D_i$ $(i=1,2,\ldots,n)$. Each C_i is oriented as boundary of D_i. We introduce a Riemannian structure on M, with area element σ, Gaussian curvature K, a curvilinear coordinate s_i on C_i, geodesic curvature k_i of C_i, and angle α_i between the tangent of C_i and the vector X.

By remark iii) of A.10 we obtain

$$\int_V K\sigma = \int_V d\varepsilon = \sum_{i=1}^{n}\left(\int_{C_i} k_i ds_i - \int_{C_i} d\alpha_i\right).$$

Similarly let Y_i be a vector field without singularity on D_i, and β_i the angle between the tangent of C_i and Y_i; then

$$\int_{D_i} K\sigma = \int_{C_i} d\beta_i - \int_{C_i} k_i ds_i .$$

Together with the theorem of Gauss-Bonnet this leads to

$$2\pi\chi(M) = \sum_{i=1}^{n}\left(\int_{C_i} d\beta_i - \int_{C_i} d\alpha_i\right) .$$

The remainder of the proof is a consequence of the following lemma.

Q.E.D.

3.2. LEMMA. Let X denote a differentiable vector field on $\mathbb{R}^2$ having no singular point on $\mathbb{R}^2 - \{0\}$, and let for an arbitrary Riemann structure on $\mathbb{R}$ denote by α the angle between the tangent of the circle S^1 and X. Then the index of X at 0 may be expressed as $1 - \frac{1}{2\pi}\int_{S^1} d\alpha$.

Proof. If β is the angle between X and the tangent of S^1 with respect to the Euclidean structure of $\mathbb{R}^2$, then

$$\int_{S^1} d\alpha = \int_{S^1} d\beta \ ,$$

since the angles α and β differ by less than π. We may then interpret β also as the difference between the angle of the tangent of S^1 and the x-axis and the angle between X and the x-axis; hence

$$\int_{S^1} d\beta = 2\pi - 2\pi i_X(0) . \qquad \text{Q.E.D.}$$

3.3. Exercise. Let M be a compact surface without boundary carrying a Riemannian structure, and let X be the gradient field of a smooth function f on M. Since the index of a gradient field at a singular point is independent of the choice of Riemannian metric we deduce: if the critical points of f are non-degenerate then the index of a singular point of X is either +1 or -1 according to whether the corresponding critical point of f is an extremum or a saddle point. Letting therefore denote by a,b, and c the number respectively of minima, saddle points, and maxima of f, we arrive at the relation of Morse: $a-b+c = \chi(M)$.

In particular for an orientable (respectively non-orientable) surface of genus g the number of saddle points of f is at least 2g (respectively g+1).

3.4. Exercise. Let M be a compact surface with boundary. Then the index theorem remains valid for smooth vector fields on M which are transverse to the boundary and have a finite number of singular points.

APPENDIX: ELEMENTS OF DIFFERENTIAL GEOMETRY OF SURFACES

We treat this subject here by means of the "moving frame" of Elie Cartan.

Let M be an <u>oriented</u> surface, provided with a Riemannian structure, with or without boundary.

A.1. DEFINITION. A <u>moving frame</u> on an open set U of M is an ordered pair (e_1, e_2) of vector fields on U forming at each point x of U an orthonormal, positively oriented base of T_xU.

We let then $(\varepsilon_1, \varepsilon_2)$ denote its moving co-frame, i.e. the ordered pair of Pfaffian forms on U characterized by $\varepsilon_i(e_j) = \delta_{ij}$.

If $(\hat{e}_1, \hat{e}_2)$ is a second moving frame on U, and $x \in U$ a variable point, then there is a map $\tau: U \to S^1$ such that

$$\hat{e}_1 = \cos\tau(x) \cdot e_1 + \sin\tau(x) \cdot e_2 ,$$

$$\hat{e}_2 = -\sin\tau(x) \cdot e_1 + \cos\tau(x) \cdot e_2$$

Correspondingly we have

$$\hat{\varepsilon}_1 = \cos\tau(x) \cdot \varepsilon_1 + \sin\tau(x) \cdot \varepsilon_2 ,$$

$$\hat{\varepsilon}_2 = -\sin\tau(x) \cdot \varepsilon_1 + \cos\tau(x) \cdot \varepsilon_2 ;$$

hence

$$\hat{\varepsilon}_1 \wedge \hat{\varepsilon}_2 = \varepsilon_1 \wedge \varepsilon_2 .$$

Thus we have shown:

A.2. <u>PROPOSITION</u>. There exists a uniquely determined volume form σ on M which induces the form $\varepsilon_1 \wedge \varepsilon_2$ on every open set U of M which carries a moving frame (e_1,e_2).

σ is called the (oriented) <u>area element</u> of the Riemannian structure. If M is compact the quantity $A = \int_M \sigma$ is called its <u>area</u>.

A.3. <u>LEMMA</u>. Let (e_1,e_2) be a moving frame on an open set U, and let $(\varepsilon_1,\varepsilon_2)$ be its moving co-frame. Then there is a uniquely determined Pfaffian form ε on U satisfying

$$d\varepsilon_1 = \varepsilon_2 \wedge \varepsilon \quad \text{and} \quad d\varepsilon_2 = -\varepsilon_1 \wedge \varepsilon .$$

Indeed we have

$$\varepsilon = \varepsilon_1([e_1,e_2])\varepsilon_1 + \varepsilon_2([e_1,e_2])\varepsilon_2 .$$

Moreover if $(\hat{e}_1,\hat{e}_2)$ is a second moving frame on U and $\hat{\varepsilon}$ their corresponding Pfaffian form then we have $\hat{\varepsilon} = \varepsilon - d\tau$.

Hence we have

A.4. <u>THEOREM</u> <u>(theorema egregium of Gauss)</u>. There exists a differentiable function K on M such that

$$d\varepsilon = K\sigma$$

holds on every open set U where a moving frame (e_1,e_2) is defined.

K is called the <u>Gaussian curvature</u> of the Riemannian structure.

A.5. <u>The unit tangent bundle</u>. Let E denote the set of all triples $(m;v_1,v_2)$ where m is a point of M and (v_1,v_2) is a positively oriented orthonormal base of $T_m(M)$; moreover let p be the projection $(m;\, v_1,v_2) \mapsto m$ from E onto M.

Let (e_1,e_2) be a moving frame on an open set U of M. Then the map $h: (m,\theta) \longmapsto (m; \cos\theta\, e_1(m) + \sin\theta\, e_2(m), -\sin\theta\, e_1(m) + \cos\theta\, e_2(m))$ is a bijection from $U \times S^1$ onto $p^{-1}(U)$ which is compatible with the projections onto U. Furthermore let $\hat{h}: U \times S^1 \longrightarrow p^{-1}(U)$ be the analoguous bijection corresponding to a second moving frame $(\hat{e}_1, \hat{e}_2)$ on U. Then we have $\hat{h}^{-1}(h(m,\theta)) = (m, \theta - \tau)$.

We provide now the set E with the uniquely determined differentiable structure for which the <u>charts</u> h become diffeomorphisms. The projection p then will be differentiable, and $p: E \longrightarrow M$ is called the <u>unit tangent bundle</u> on M (actually here the bundle of orthonormal positively oriented bases of the tangent spaces of M).

The group S^1 acts differentiably on E by the operations $R_\alpha: (m; v_1, v_2) \longmapsto (m; \cos\alpha\, v_1 + \sin\alpha\, v_2, -\sin\alpha\, v_1 + \cos\alpha\, v_2)$; if h is a chart on E then we have

$$h^{-1}(R_\alpha(h(m,\theta))) = (m, \theta+\alpha).$$

A.6. <u>LEMMA</u>. Let $\omega_1, \omega_2: T(E) \longrightarrow \mathbb{R}$ be maps defined on $T_{(m;v_1,v_2)}(E)$ by $\langle v_1, p^T w\rangle$; respectively $\langle v_2, p^T w\rangle$. Then ω_1 and ω_2 are Pfaffian forms on E.

Indeed letting $h: U \times S^1 \longrightarrow p^{-1}(U)$ denote a chart of E corresponding to a moving frame (e_1, e_2) we find

$$h^*\omega_1 = \cos\theta\, \varepsilon_1 + \sin\theta\, \varepsilon_2 \quad \text{and} \quad h^*\omega_2 = -\sin\theta\, \varepsilon_1 + \cos\theta\, \varepsilon_2 .$$

These expressions show also that the forms ω_1 and ω_2 have the following properties:

i) $\omega_1 \wedge \omega_2 = p^*\sigma$;

ii) $R_\alpha^*\omega_1 = \cos\alpha\, \omega_1 + \sin\alpha\, \omega_2$;

iii) $R_\alpha^*\omega_2 = -\sin\alpha\, \omega_1 + \cos\alpha\, \omega_2$.

A.7. <u>LEMMA</u>. There exists a uniquely determined Pfaffian form ω on E such that $d\omega_1 = \omega_2 \wedge \omega$ and $d\omega_2 = -\omega_1 \wedge \omega$.

Indeed for a chart $h: U \times S^1 \to p^{-1}(U)$ of E corresponding to a moving frame (e_1, e_2) we have

$$h^*(d\omega_1) = (h^*\omega_2) \wedge (\varepsilon - d\theta)$$

and

$$h^*(d\omega_2) = -(h^*\omega_1) \wedge (\varepsilon - d\theta) \; .$$

For another chart $\hat{h}: U \times S^1 \to p^{-1}(U)$ belonging to a second moving frame $(\hat{e}_1, \hat{e}_2)$ we find the relation

$$(\hat{h}^{-1} \circ h)^*(\hat{\varepsilon} - d\hat{\theta}) = \varepsilon - d\theta \; .$$

We call ω (or $-\omega$) the <u>connection form</u> of the Riemannian structure (cf. remark ii) of A.10).

A.8. <u>LEMMA</u>. We have $d\omega = p^*(K\sigma) = (K \circ p)\omega_1 \wedge \omega_2$.

A.9. <u>PROPOSITION</u>. Let M be a compact and parallelizable surface without boundary. Then

$$\int_M K\sigma = 0 \; .$$

Indeed, the unit fibre bundle then has a section $s: M \to E$, and therefore

$$\int_M K\sigma = \int_M (p \circ s)^* K\sigma = \int_M s^* d\omega = 0 \; .$$

A.10. <u>Remarks</u>.

i) The form $\Omega = \omega_1 \wedge \omega_2 \wedge \omega$ is a volume form on E: if $h: U \times S^1 \to p^{-1}(U)$ is a chart of E corresponding to a moving frame (e_1, e_2) then $h^*\Omega = -\sigma \wedge d\theta$. Hence for a compact M

$$\int_E \Omega = -2\pi A \; .$$

The three Pfaffian forms ω_1, ω_2, and ω determine therefore a parallelism on E.

ii) Let X be the vector field on E which generates the one-parameter group R_α. Then

$$\omega(X) = -1, \quad L_X\omega_1 = \omega_2, \quad L_X\omega_2 = -\omega_1, \quad L_X\omega = 0 .$$

In particular the distribution of planes defined by the form ω on E is transverse to every fiber of p and invariant under the action of S^1. It is integrable if and only if the Gaussian curvature vanishes.

iii) Let c denote a regular curve in M, and let γ be the curve in E corresponding to the Frenet frame (τ,ν) of c. The curve γ is a lift of c (i.e. $p \circ \gamma = c$), and we have $\gamma^*\omega_1 = \|c'(t)\|dt$ and $\gamma^*\omega_2 = 0$. We define a curvilinear coordinate (arc length) of c as an indefinite integral s of $\gamma^*\omega_1$, and we call the function k(s) satisfying $\gamma^*\omega = -k(s)\,ds$ the <u>geodesic curvature</u> of c.

Moreover let c lie in an open set U of M on which there is defined a moving frame (e_1, e_2), and let $h: U \times S^1 \to p^{-1}(U)$ be the corresponding chart of E. Then $\alpha = p_2 \circ h^{-1} \circ \gamma$ is called the "angle" between the tangent τ of c and e_1.

We then have $\gamma^*\omega = -k(s)ds = c^*\varepsilon - d\alpha$.

iv) Let γ be a regular curve in E satisfying $\gamma^*\omega = 0$. Then we say that γ defines a <u>parallel transport</u> of frames along the curve $c = p \circ \gamma$ in M (c is then a regular curve as well).

v) A regular curve c in M is called a <u>geodesic</u> if it has constant speed $\|c'\|$, and if its family of Frenet frames is parallel along c, or if its geodesic curvature vanishes identically.

The vector field Y on E characterised by the relations

$$\omega_1(Y) = 1 \qquad \text{and} \qquad \omega_2(Y) = \omega(Y) = 0$$

is called the <u>geodesic spray</u> on E. The projection $p: E \to M$ establishes a 1:1 correspondence between its integral curves and the geodesics of speed 1 on M. It follows in particular that there is a unique geodesic of M through a given point with a given tangent.

vi) Let f be a direct isometry of M (i.e. a diffeomorphism of M preserving the orientation and the Riemannian structure). Then the map $F:(m;\, v_1, v_2) \longmapsto (f(m);\, f^T v_1, f^T v_2)$ is an automorphism of the tangent bundle E which leaves invariant each of the forms ω_1, ω_2, and ω, as well as the geodesic spray Y. Hence f transforms every geodesic of M into a geodesic.

A.11 <u>THEOREM (theorem of Gauss-Bonnet)</u>. Let M be a compact surface without boundary. Then

$$\int_M K\sigma = 2\pi\chi(M) \quad .$$

<u>Proof</u>.

i) Let $M = S^2$. Then it may be written as the union of two hemispheres D_1 and D_2 with a great circle C as their common boundary which we orient as the boundary of D_1. Let (e_1, e_2) be a moving frame on an open neighbourhood of D_1, and denote by α the angle between the tangent to C and e_1 and by k the geodesic curvature of C. By lemma 3.2 we find:

$$\int_{D_1} K\sigma = \int_{D_1} d\varepsilon = \int_C \varepsilon = \int_C d\alpha - \int_C k(s)\, ds = 2\pi - \int_C k(s)\, ds \ .$$

Similarly

$$\int_{D_2} K\sigma = 2\pi + \int_C k(s)\, ds \quad ,$$

and therefore

$$\int_{S^2} K\sigma = 4\pi = 2\pi\chi(S^2) \ .$$

ii) Let $M = T^2$. By proposition A.9 we have

$$0 = \int_{T^2} K\sigma = \chi(T^2).$$

iii) Let V denote the complement with respect to T^2 of two disjoint open discs D_1 and D_2 with disjoint Jordan curves C_1 and C_2 as oriented boundary. Moreover we choose a moving frame (e_1, e_2) on an open neighbourhood of V, and denote by α_1 (respectively by α_2) the angle of the tangent of C_1 (respectively of C_2) with e_1. If k_i is the geodesic curvature of C_i (i=1,2) we obtain as above

$$\int_V K\sigma = -\int_{C_1} d\alpha_1 + \int_{C_1} k_1(s_1)ds_1 - \int_{C_2} d\alpha_2 + \int_{C_2} k_2(s_2)ds_2 .$$

Making use of the Gauss-Bonnet formula for T^2, this reduces to

$$\int_V K\sigma = -4\pi + \int_{C_1} k_1(s_1)ds_1 + \int_{C_2} k_2(s_2)ds_2 .$$

If M is a surface of genus $g \geqslant 2$ then it may be represented as the union of g+2 compact submanifolds with boundary $V_o, V_1, \ldots, V_g, V_{g+1}$ with the following properties:

- V_o (respectively V_{g+1}) is diffeomorphic to the disc D^2 and has a Jordan curve C_o' (respectively C_{g+1}) as its boundary;
- V_i, $i=1,\ldots,g$ is diffeomorphic to the manifold V considered above, and it has two Jordan curves C_i and C_i' as its boundary;
- $C_i' = C_{i+1}$ for $i = 0,1,\ldots,g$.

Hence we deduce

$$\int_M K\sigma = \sum_{i=0}^{g+1} \int_{V_i} K\sigma = 4\pi - 4\pi g = 2\pi\chi(M). \qquad \text{Q.E.D.}$$

A.12. Exercise. The half-plane

$$H = \{z = x + iy \in \mathbb{C} \mid y > 0\}$$

provided with the metric

$$\frac{dx^2 + dy^2}{y^2} = -4 \frac{dz\, d\bar{z}}{(z - \bar{z})^2}$$

is called <u>Poincaré's half-plane</u> or hyperbolic half-plane.

i) The angles, as defined by this Riemannian structure coincide with those of the Euclidean structure.

ii) The Gaussian curvature of H is constant and equals -1.

iii) The geodesics are the intersections of H with the vertical straight lines and the circles having their centres on the real axis. Hence there is a unique geodesic joining any two distinct points of H.

iv) Let P denote a geodesic polygon in H (i.e. a polygon with geodesic arcs as sides). Then

$$\iint_P \sigma = \sum_i \alpha_i - 2\pi ,$$

where the α_i are the exterior angles of P.

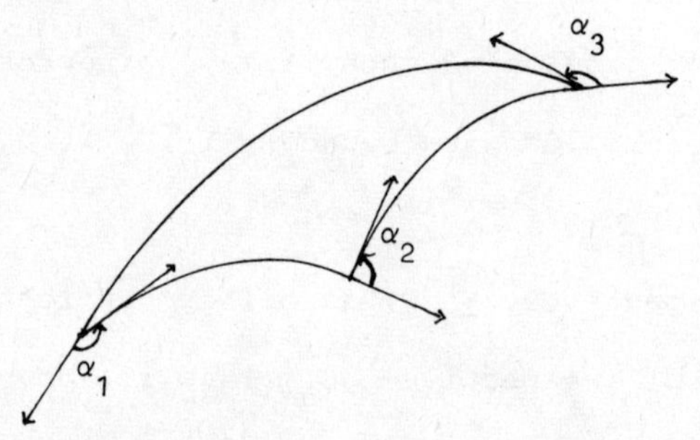

In particular the sum of the angles of a "triangle" in H is strictly less than π, and its area equals its angular defect, i.e. the amount by which π exceeds the sum of the angles.

v) The real homographies $z \mapsto \frac{az+b}{cz+d}$, $ad - bc > 0$, leave H invariant and preserve its Riemannian structure. Hence the group $PSl(2,\mathbb{R}) = Sl(2,\mathbb{R})/\{\pm I\}$ may be identified with the group of proper isometries of H. The corresponding action of $PSl(2,\mathbb{R})$ on the bundle of unit tangent vectors of H is transitive and free.

vi) Let (z_1, z_2) and (Z_1, Z_2) denote two distinct pairs of points of H with equal distance. Then there is a unique proper isometry of H carrying z_1 into Z_1 and z_2 into Z_2.

vii) The Pfaffian form $\Omega = \omega_2 + \omega$ on the unit tangent bundle E is exact. It defines a foliation $\mathcal{F}$ of codimension 1 of E which is "invariant under the isometries of H".

The leaves of $\mathcal{F}$ consist of the Frenet frames of the geodesics of H which have the same "α-limit point". Note that the map $t \longmapsto (t, 1, -\frac{\pi}{2})$ of $\mathbb{R}$ into $H \times S^1$ determines an integral curve of Ω, and every leaf of $\mathcal{F}$ is invariant under the one-parameter group generated by the geodesic spray.

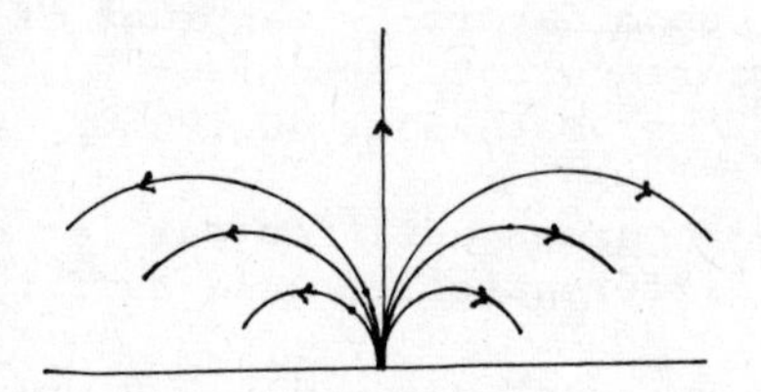

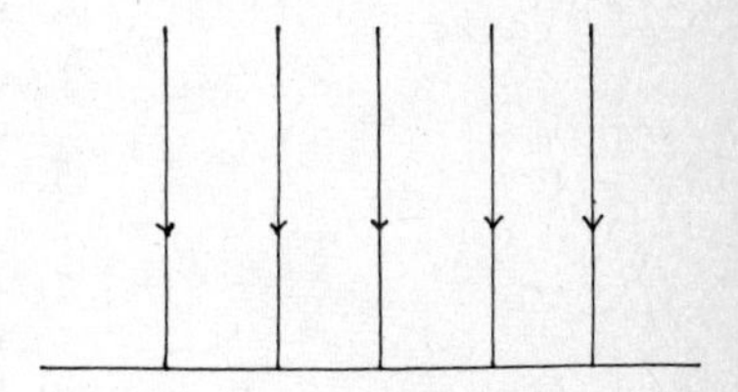

Bibliography

H. Poincaré: Sur les courbes définies par une équation différentielle.
J.Math. Pures et Appl., 1881,pp.375-425.
J.Math. Pures et Appl., 1882,pp.251-296.
J.Math. Pures et Appl., 1885,pp.167-244
(and Oeuvres complètes, Gauthiers-Villars, Paris, vol.I, 1928).

---

[1] V.I. Arnold: Small Denominators I. Mappings of the Circumference onto Itself.
Transl.Amer.Math.Soc.,46,1965,pp.213-284.

[2] A. Denjoy: Sur les courbes définies par les équations différentielles à la surface du tore.
J.Math.Pures et Appl.,11,1932,pp.333-375.

[3] H. Furstenberg: Strict Ergodicity and Transformations of the Torus.
Amer.J. of Math.,83,1961,pp.573-601.

[4] A. Haefliger - G. Reeb: Variétés (non séparées) à une dimension et structures feuilletées du plan.
Ens.Math.,3,1957, pp.107-126.

[5] M. Herman: Sur la conjugaison différentiable des difféomorphismes du cercle à des rotations.
Publ.Math. I.H.E.S., 49, 1979, pp.5-233.

[6] H. Kneser: Regulaere Kurvenscharen auf den Ringflaechen.
Math.Ann.,91,1923, pp.135-154.

[7] N. Kopell: Commuting Diffeomorphisms.
Proc.Symp. Pure Math.,14,1970, pp.165-184.

[8] A.J. Schwartz: A Generalization of a Poincaré-Bendixson Theorem to Closed Two Dimensional Manifolds.
Amer.J. of Math.,85,1963, pp.453-458

[9] S. Sternberg: Local C^n-Transformations of the Real Line.
Duke Math.J.,24,1957, pp.97-102.